Palgrave Historical Studies in t
and its Afterli

Series Editors
Owen Davies
School of Humanities
University of Hertfordshire
Hatfield, UK

Elizabeth T. Hurren
School of Historical Studies
University of Leicester
Leicester, UK

Sarah Tarlow
School of Archaeology and Ancient History
University of Leicester
Leicester, UK

This limited, finite series is based on the substantive outputs from a major, multi-disciplinary research project funded by the Wellcome Trust, investigating the meanings, treatment, and uses of the criminal corpse in Britain. It is a vehicle for methodological and substantive advances in approaches to the wider history of the body. Focussing on the period between the late seventeenth and the mid-nineteenth centuries as a crucial period in the formation and transformation of beliefs about the body, the series explores how the criminal body had a prominent presence in popular culture as well as science, civic life and medico-legal activity. It is historically significant as the site of overlapping and sometimes contradictory understandings between scientific anatomy, criminal justice, popular medicine, and social geography.

More information about this series at
http://www.palgrave.com/gp/series/14694

Sarah Tarlow · Emma Battell Lowman

Harnessing the Power of the Criminal Corpse

Sarah Tarlow
School of Archaeology and Ancient
    History
University of Leicester
Leicester, UK

Emma Battell Lowman
School of Humanities
University of Hertfordshire
Hertford, UK

Palgrave Historical Studies in the Criminal Corpse and its Afterlife
ISBN 978-3-030-08570-4   ISBN 978-3-319-77908-9   (eBook)
https://doi.org/10.1007/978-3-319-77908-9

Cover illustration: Giovanni Aldini (1762–1834), a galvanism experiment on a corpse.
Courtesy of the Wellcome Collection

Printed on acid-free paper

This Palgrave Macmillan imprint is published by the registered company Springer
International Publishing AG part of Springer Nature
The registered company address is: Gewerbestrasse 11, 6330 Cham, Switzerland

Acknowledgements

This book is the summation of the research programme 'Harnessing the Power of the Criminal Corpse', funded by the Wellcome Trust (Grant No. 095904/Z/11/Z), and we are grateful to the Trust for their intellectual, logistical and financial support at every stage of the project. Our thanks go to our colleagues on the project: Rachel Bennett, Owen Davies, Zoe Dyndor, Elizabeth Hurren, Peter King, Patrick Low, Francesca Matteoni, Shane McCorristine, Floris Tomasini, Richard Ward, and Clare Canning. Thanks also to our institutions, the University of Leicester and the University of Hertfordshire, for supporting the project. The gestation of this book has been longer than we had hoped because its production was severely affected by the illness and death of ST's husband, Mark Pluciennik. ST would like to thank all her colleagues and friends for their patience and support over this difficult time. EBL in particular thanks Adam Barker for his constant practical and emotional support.

CONTENTS

LIST OF FIGURES

The Criminal Corpse in History

CHAPTER 1

Introduction

This book is about the power of the dead body. This power is rooted in paradox: deprived of will and the capacity to take action, to think, speak, coerce or persuade, deprived of life itself, the human body can still be a powerful agent of change.

On 3 September 2015 news media around the world carried a photograph distributed by a Turkish news agency of the tiny body of 3-year-old Alan Kurdi washed up on a beach. The little boy, dressed in a red T-shirt and blue shorts, drowned when the boat carrying him and his family from Turkey to the nearby Greek island of Lesbos sank. The photograph was released at the height of the European migrant crisis of 2015. Hundreds of thousands of people, many from Syria but also substantial numbers from Iraq, Afghanistan, and sub-Saharan Africa were travelling to Europe, many in extremely dangerous ways due to their desperation to reach safety. Five days before Alan Kurdi lost his life the number of migrants drowned in the Mediterranean during this crisis was already over 2400.[1]

The European reaction to these migrants—both political and popular—had not been positive. Many European leaders reacted by trying to strengthen their borders, police the Mediterranean, and institute rigorous and deterrent immigration policies. But the publication of the Alan Kurdi photographs proved a salutary and transformative moment. In the United Kingdom, the British Prime Minister David Cameron changed his rhetoric from how best to protect the nation from the

© The Author(s) 2018   3
S. Tarlow and E. Battell Lowman, *Harnessing the Power of the Criminal Corpse*, Palgrave Historical Studies in the Criminal Corpse and its Afterlife, https://doi.org/10.1007/978-3-319-77908-9_1

'swarm' of migrants to announcing new, much higher numbers of Syrian refugees that would be taken in by the United Kingdom.[2] Marches in London, Vienna and elsewhere mobilised popular support and sympathy for the human plight and suffering of the migrants. Crowds gathered at German stations to welcome new arrivals with gifts of food, toys, and clothes. It would be an exaggeration to claim that one picture turned the tide of political opinion, but it is fair to say that the unutterably tragic and affecting photograph of a drowned toddler on a Turkish beach made more of a difference than the raw numbers of drowned migrants that had featured in almost every news broadcast of the previous six months.[3] Alan Kurdi was far from the only child to die in this crisis, but the image of his tiny corpse struck a nerve with viewers across Europe and around the world.

The remarkable response to the images of Alan Kurdi's corpse illustrates the emotive power of a dead body. In death, although he was a member of a group widely reviled and feared in Europe, he was named and individualised. In the image of his lifeless corpse, many commentators found a connection: he 'could have been my child'.

There is universality to the corporeal nature of being human. Across a vast range of experiences, what we have in common is that we each inhabit a body and that at some point it will die. It is easy to see how the living body is a powerful, active and agentive thing. As we write, as you read, our bodies are engaged and critical to the experience of creation and relation taking place via these pages. But a dead body is surely an utterly different kind of thing. Insensibly inanimate, the corpse is not a person as we usually understand it, but a thing, an object, or even what Julia Kristeva calls an 'abject'.[4] How can a thing, incapable of independent movement, thought or utterance, be active? This question, usually applied to material artefacts, or particular arrangements of space, is an essential one for theorists of archaeology, and one which has been addressed with great sophistication over the last 30 years, notably by Ian Hodder and the post-processual school of archaeological theory.[5]

Hodder claimed that the traditional archaeological interpretation that material culture passively signals or records past events, intentions and sociocultural realities is flawed. Instead, he argued, material culture is actively involved in structuring social reality.[6] Thus for example, wearing high-heeled strappy shoes is not (only) a way of signalling or codifying womanhood in contemporary society, but high heels are actively part of what creates a woman in particular cultural contexts.

This means that particular forms of organised space, technology, clothes or objects actively create situations and relationships. Whether we serve tea or coffee in a fine china cup and saucer, as part of a matching set, or in a big, chunky, earthenware mug works to structure an encounter and constitute an identity; it gives the thing a certain active power.

Archaeologists have not usually included the dead body in their discussions of the agentive power of things, despite cogent arguments that the body can and often should be regarded as material culture.[7] Like a ceramic vessel or a flint axe, the body is shaped by technologies that can thicken a bone, shape a skull or remodel the flesh. And like a pot or an axe, bodies bear the marks of use, stress, repair, and ornamentation. If the body is a thing, then it is sometimes useful to understand it as an active thing.

This book, then, proceeds from an understanding that dead bodies can be powerful and can, indeed, play significant roles in the negotiation of social and cultural beliefs. We also take the position that there is no contradiction between a dead body that is powerful and one that is manipulable, although there are constraints on how far a body can be manipulated, and the possibility of unintended or subversive meanings emerging from them cannot be entirely controlled.

THE CRIMINAL CORPSE

Our study is of a particular kind of body—the criminal corpse—and of how its power was harnessed, channelled and sometimes subverted or resisted to bring about a multitude of intended and unintended ends. In particular, we address three key concerns: to place the history of legislated post-mortem punishment in eighteenth- and nineteenth-century Britain into the *longue durée*; to complete the journey of the condemned which until now has ended at the gallows by following the corpse from execution through to the conclusion of mandated post-mortem punishments; and to identify the ways the power of these criminal corpses has been harnessed in Britain including up to the present day.

Our attempt to produce a comprehensive response to these questions is rooted in the findings of the 5-year Wellcome Trust-funded project 'Harnessing the Power of the Criminal Corpse'. The project arose from Sarah Tarlow's long-term concern with the nature of beliefs about the dead body in early modern Britain and Ireland.[8] In previous work, her key question 'What did people really believe about the dead body?'

turned out to be unanswerable. She found that beliefs in early modern theological discourse were quite different to beliefs about the medical and scientific body. Both of these kinds of beliefs were incompatible with those evident in actual material practice and by popular accounts or folklore. Rather than the result of doctors having different beliefs from vicars, or the common people having different beliefs from the educated, these multiple discourses existed in parallel and any single person might draw on one set of beliefs or another, dependent on context.

Beliefs about the dead body in Britain were (and are today) ill-defined, multiple and often incommensurable. Theological writings in the early modern period insisted that the dead body was meaningless, and that what mattered was the soul. Contrary to what is sometimes asserted, no early modern theologian appears to have claimed that a whole body was necessary for bodily resurrection in the Christian tradition: quite the reverse. Similarly, anatomical and medical science at the time was developing a paradigm of the body as a machine of predictably interacting systems. Furthermore, and intriguingly, Tarlow found that the complex polyvalence of beliefs about the dead body was especially evident in the case of the executed criminal body and that attitudes to execution and the executed body in particular highlighted these incompatibilities.[9]

For this reason, Tarlow put the criminal body, around which numerous traditions of discourse spin and have spun out, at the centre of the multidisciplinary 'Harnessing the Power of the Criminal Corpse' project. Led by Tarlow in archaeology, the structure and scope of the project was developed and managed in close collaboration with Owen Davies in folklore, Peter King in legal history, and Elizabeth Hurren in medical history. The project focusses on the post-mortem treatment of criminals who were executed, mainly in Britain, between the mid-eighteenth and the mid-nineteenth centuries.

These dates demarcate the core period of the study because of the impact of the Murder Act. In 1752 a new act came into force, aimed at marking out those convicted of murder for particular judicial censure. The Act states:

> [W]hereas the horrid Crime of Murder has of late been more frequently perpetrated than formerly... And whereas it is thereby become necessary that some further Terror and peculiar Mark of Infamy be added to the Punishment of Death, now by Law inflicted on such as shall be guilty of the said heinous Offence... Sentence shall be pronounced in open Court,

immediately after the Conviction of such Murderer… in which Sentence shall be expressed, not only the usual Judgment of Death, but also the Time appointed for the Execution thereof, and the Marks of Infamy hereby directed for such Offenders, in order to impress a just Horror in the Mind of such Offender, and on the Minds of such as shall be present, of the heinous Crime of Murder.

And after Sentence is pronounced, it shall be in the Power of any such Judge, or Justice, to appoint the Body of any such Criminal to be hung in Chains; but that in no Case whatsoever, the Body of any Murderer shall be suffered to be buried, unless after such Body shall have been dissected and anatomized.[10]

In practice this usually meant that a judge sentencing a murderer would specify that following execution the criminal's body should be sent to an appointed surgeon or anatomist for dissection, or turned over to the sheriff to be hung in chains ('gibbeted').

The Murder Act remained in force until it was superseded by the Anatomy Act of 1832. Under that Act, inspired by the convergence of the growing need of medical scientists for a more secure and plentiful supply of cadavers for dissection, the public outcry over grave robbing scandals, and the notorious Burke and Hare murders of the 1820s (in which sixteen people on the fringes of society were murdered in order for their bodies to be sold to Edinburgh's anatomists), the bodies of the 'unclaimed' poor from workhouses and hospitals replaced the bodies of criminals as the main legal source of dissection material.[11] The post-mortem punishment of gibbeting ended in 1832 and was taken off the books two years later in response to changing sensibilities and ideas about punishment.

The project team assembled to accomplish the goal of examining the shifting power of the criminal corpse centred on the Murder Act, and the broader implications for Britons then and now, included archaeologists, historians, folklorists, philosophers and sociologists, and proceeded along six distinct research themes, each of which used the disciplinary tools best suited to understanding and investigating the questions arising in that area. Each strand produced original research and published their findings in the forms of monographs (books) and articles. Importantly, these outputs are available online for free thanks to the financial support of the Wellcome Trust and in accordance with their mandate to make research public.

In the first research strand, *The Criminal Justice System and the Criminal Corpse*, historians Peter King and Richard Ward investigated the debates and historical context in which the Murder Act was created. This strand paid particular attention to the role of print culture and the use of pre-execution aggravated punishment in Europe in the development of the British government's legislative response to the perceived problem of rising murder rates in mid-eighteenth century England. King and Ward's work not only addressed the specific history of the emergence and construction of the Murder Act, but also of how the Act worked in practice. Close attention to the relationship between court records and the expense claims that indicate which punishments were actually carried out under the Act shed new light on regional differences in rates of conviction and post-mortem punishment and in so doing drew attention to the discretion exercised by law-keepers during the life of the Act. The work of King and Ward was key to understanding the legislative and legal processes by which criminal corpses destined for formalised post-mortem punishment were created in the eighteenth and nineteenth centuries in Britain.[12]

Historian of medicine Elizabeth Hurren carried out the second research strand, *The Criminal Corpse in the Expanding Anatomical and Medical World of Georgian Society*. This work focused on the first of two post-mortem punishments mandated under the Murder Act: 'anatomisation and dissection'. Hurren examined not only the medical procedures this punishment involved, but also the competing interests of the execution crowd, the state, and the medical men for whom legal access to criminal corpses represented valuable possibilities in terms of research, profit, and renown. Further, this strand involved groundbreaking work in understanding the role of the crowd and of display spaces involved in this spectacular form of punishment.[13]

Sarah Tarlow and Zoë Dyndor traced the journey of the criminal corpses into forms of display, particularly on the gibbet, in the strand *Placing the Criminal Corpse*. This strand involved not only a full survey of the whole and partial gibbet cages that exist today in Britain, but also a spatial survey and analysis of the practice of gibbeting. The analysis of extant cages laid alongside investigation of the expense claims made by Sheriffs for the construction of gibbets made possible new and detailed understandings of this previously understudied and peculiarly British form of punishment. Tarlow and Dyndor traced the journey of the criminal corpse from the gallows to the gibbet and into the days, months, years, or decades suspended between earth and sky during which its significance and meaning shifted and slipped from deterrent display to macabre spectacle to mundane marker on the landscape.[14]

It is not only as a cultural signifier that the criminal corpse had power in the world of the Murder Act. The very substance of the body had power. In some ways not fully understood or articulated by contemporaries, a vigorous body suddenly cut off—which was the case with most executed criminals who were overwhelmingly young adult men—was believed to have some residual life force which could be channelled for the benefit of the living. This involved physical contact or even ingestion of the material body itself. In *The Dead Sustaining Life: Criminal Corpses in European Medicine and Magic, 1700–1900*, Owen Davies and Francesca Matteoni took a folkloric approach to investigate how criminal corpses—both those created under the Murder Act in Britain but also criminal corpses more broadly—were used for medicinal and magical purposes. This strand went beyond the Murder Act both in temporal scope and also geographically to investigate the varied uses of criminal corpses in over two hundred years in Europe (including Britain). In this strand, Davies and Matteoni traced how the criminal corpse journeyed, usually in pieces, into new contexts and in so doing, how it acquired new meaning and potency.[15]

Shane McCorristine followed the criminal corpse into narrative, in particular literary fiction in eighteenth- and nineteenth-century Britain. *The Criminal Corpse in Pieces* considered the way fiction writing of this period is replete with criminal bodies in the form of punished corpses, vengeful ghosts, and powerful relics. The way these stories were constructed, told, and read reveals a remarkably wide engagement between the public and the criminal corpse, one far beyond the thousands who participated in the spectacles of post-mortem punishment directly, for example in the anatomisation displays described by Hurren or the carnival crowd that attended gibbets in the work of Tarlow and Dyndor.[16]

In the final strand, *The Criminal Corpse Remembered: Historical and Contemporary Perspectives on Power, Agency, Values and Ethics*, philosopher Floris Tomasini engaged with the ethical legacies of the Murder Act and the Anatomy Act, the legislation that replaced criminal corpses with those of the poor and unclaimed for the use of medical science. Issues of ownership and use around criminal bodies are thus tied to a longer trajectory of the use of corpses for purposes other than those involved in end-of-life ceremonies. Tomasini calls for greater historical awareness of developing cultural attitudes towards the proper treatment of the dead body. A deeper time context enables a more sophisticated approach to contemporary ethical debates.[17]

Work within the six strands proceeded according to the disciplinary parameters and protocols of those directly involved, but was informed by the work taking place in the other strands. Throughout the life of the project, the teams met, exchanged draft work, and discussed their ideas and findings. These regular project meetings facilitated multidisciplinary discussions at all stages of the research process and supported the effective sharing of evidence, sources, and techniques. The statistics compiled by Ward and King for their crime history work provided the framework for Hurren's medical history and Tarlow's gibbet study. A 2013 discussion of the dissection of William Corder sent McCorristine onto a new research trajectory, resulting in his monograph on the popular reception of this notorious case.[18] Though critical to the success of the project, these discussions were not necessarily easy. The criteria and evidential basis required to make a claim or conclusion in the history of law are very different from, say, those required to make a claim or conclusion in archaeology. In this case, not only the types of evidence differ (textual versus material), but the quantity of evidence required to substantiate a conclusion or claim differs between disciplines. Where archaeologists are generally happy—indeed obliged—to make a case on the basis of very scanty evidence and a plausible hypothesis, historians of law in modern Britain are used to having huge, statistically robust databases capable of demonstrating quite subtle chronological and geographical patterning. Project members had to come to terms with seeing from a new perspective the limitations and benefits of our respective disciplines, and maintaining our communication and collaboration, both despite and in relationship with, our divergent practices. The publications and findings of the individual strands clearly show, however, that these deeper epistemological challenges can result in exciting and illuminating results.

This book is the capstone to the 'Harnessing the Power of the Criminal Corpse' project. It brings together the research and findings from across the project to create an intentionally interdisciplinary work that is at once relevant and useful to the disciplines involved in its production, but also responds to broader questions that may escape the bounds of any individual scholarly approach. We have tried to make use of the project findings to build a coherent scholarly narrative on the criminal corpse focused on the period of the Murder Act that synthetically extends historical knowledge about Britain in the eighteenth and nineteenth centuries, and speaks more specifically to questions of materiality and the body, criminality, treatment of the dead and the history of punishment.

This endeavour is made possible not only by the depth and range of research conducted by the wider project team, but also by the combination of the experience of the authors, and their collaborative process. Sarah Tarlow is an archaeologist who has worked extensively on the history and archaeology of the dead body. Emma Battell Lowman is an interdisciplinary historian whose work pays particular attention to issues of power, language, the body and story. She joined the project specifically to work on final outputs and benefits from an outsider's view of the research conducted in the project strands. From the early days of work on this book, Tarlow and Battell Lowman have cultivated a close collaborative relationship that has made space for creative approaches to research and writing, and allowed all elements of the book to be truly the result of co-production and co-authorship.

To investigate the tensions around the criminal corpse during the life of the Murder Act, the book's first section seeks to put this 80-year period into a wider historical context. This *longue durée* approach begins in medieval Britain then continues into the early modern period. The culturally rich meanings of the dead criminal body in the Middle Ages and into early modernity derive in part from the symbolic resonance of the crucified Christ. The blurred line between outcast and martyr was an ongoing semiotic problem in this period.

Section II "The World of the Murder Act" concentrates on the specific historical moment of the Murder Act (1752–1832). Investigations centre on the Act in legislation and in practice, and follow the corpse into the two distinct post-mortem punishments the Act mandated, dissection and gibbeting. Though equal in the eyes of the law, these two treatments diverge sharply from one another in terms of impact, process, and legacy. This section is structured by what King has described as the journey of the criminal body: we start with the legal processes (and their socioeconomic roots) that created the particular criminal corpses at the centre of our study, the Murder Act, which takes the criminal/body from the gaol to the courtroom, to the gallows; we then continue the journey from the foot of the gallows, diverging to trace the path of those convicted murderers sentenced to dissection and those sentenced to hang in chains.

Section "Body and Power" extends the journey of the criminal/body from the sites and physical state of post-mortem punishment to the afterlives of the criminal corpses, first by following the physical remains, then by following their stories, and finally by pursuing the multifaceted legacies of the criminal corpse today. The bodies of executed criminals were not only the passive bearers of statements about political power, social deviancy, or religious orthodoxy; they were also actively involved in the creation of certain sets of power relationships, social codes, and spiritual conformities. Moreover, the discourses in which they participated were not hegemonic: dead bodies were appropriated in the construction of competing or subversive positions, as well as for socially dominant claims to power.

This volume showcases the work of the entire 'Harnessing the Power of the Criminal Corpse' team. However, it is authored by Battell Lowman and Tarlow and we have drawn selectively on colleagues' research and have not necessarily chosen to tell the histories they would tell. We have drawn on additional evidence to construct arguments which might not be the same as those which most intrigue our colleagues. Further, we write cognisant of the relevance the project findings may have for those working in other disciplines and connected areas who may not otherwise encounter the specific project strand outputs (including in particular historical sociology, historical geography and cultural history); and to respond to public interest in this subject: in Britain and beyond, people encounter legacies of the Murder Act every day—in museums, at sites of post-mortem punishment that dot the landscape, and in stories, songs and popular representations.

THE WORLD OF THE MURDER ACT

Centuries are convenient slices of time for historians. Unfortunately, the most significant cultural, political and economic changes do not always occur at regular 100 year intervals, nor do they neatly coincide with years ending in 00. Historians of the period on which we focus therefore conventionally use the term 'long eighteenth century' to define the social and political world of the time, but with enough elasticity to pull in up to half of the seventeenth century and half of the nineteenth, although its period is often slightly shorter at approximately 1688–1815 or 1832.[19] The coherence of this extended century derives from its particular importance as the location of Britain's transition to the modern.

In the areas of technology, industry, sociopolitical organisation/structure and sensibilities, Britain's long eighteenth century is, quite rightly, a critical period of study. Our temporal focus, the period from the mid-eighteenth century to the mid-nineteenth century, fits within the bounds of the long eighteenth century, though we have focused our parameters by using the years during which a particular piece of legislation was in force.

In the mid-eighteenth century, Britain was industrialising rapidly and expanding its overseas territories. New industrial and agricultural practices and structures reshaped the nation's human and physical geography. Culturally, new ideologies and codes of personal and political behaviour formed rapidly. Protestant culture underlay emerging ideologies of improvement, stressing not only the possibility of changing one's own fate and the wider world, but also the ethical imperative of doing so. New emotional codes valorised emotion and sensibility. Society was being stratigraphically reorganised from 'sorts' into classes, and movement within the system depended somewhat less on wealth alone and more on cultivation of the self and one's image. These changes all had effects on ideas about crime and punishment, disease and the body, death and mourning, and aesthetics, which directly inform the history of the criminal corpse.

In placing the period of the Murder Act into a wider historical context in order to better probe the power and significance of the criminal corpse, we not only bring together the specific findings of the different research strands of our project, we also argue for the enduring power and potency of the criminal corpse today. Our review of the centuries before the Murder Act is key to developing a *long durée* history of the criminal corpse with the potential to correct a widespread assumption about the history of punishment. Although a critique of progressivist and Whiggish historical metanarratives is nothing new, an expectation that the history of punishment follows a trajectory from more physical to more psychological, from torture to reform, from brutal to civilised, is still common. By considering more than a thousand years of capital punishment and its aftermath in Britain, even a whistle-stop and cursory examination shows that any such slow progression from savagery to civilisation is a fallacy.

Sociologist Norbert Elias famously suggested that a 'civilising process' was at work in early modern and modern European history, refining manners, softening interpersonal relationships and promoting humanitarian thinking.[20] But the history of post-mortem punishment shows that the most brutal post-mortem punishments—dissection and gibbeting—came into legal force and reached their peak popularity just at the time when the civilising process was supposed to be achieving its greatest victories: the middle of the eighteenth century.[21] This not only disrupts core understandings of broad social change in the eighteenth and nineteenth centuries, it also requires deeper investigation. This is why we have chosen to tell a story with deeper roots and an extended reach.

The core period around which our project focuses is the age of the Murder Act, from 1752 when the law first stipulated that no murderer could be buried in holy ground unless their body had first been dissected or hung in chains, to 1832, when the Anatomy Act made the Murder Act redundant. Our intent is twofold: to construct a synthesis that for the first time comprehensively follows the criminal corpse created under the Murder Act beyond the gallows, and to address a new set of questions raised by this process in combination with our own specific interests. For example, Battell Lowman's interest in the use of gibbeting in the overseas British world was prompted by her wider research into knowledge transmission in colonial contexts.

The intellectual genealogy of this book owes a great deal to the pioneering work of VAC Gatrell on the history and public reception of execution in late eighteenth- and early nineteenth-century Britain.[22] His comprehensive and at times shocking investigation into capital punishment in Britain sheds light on how executions were staged by the state and were consumed by the crowd. Gatrell's work directed our interest, and that of the project, to similar themes and concerns. In the more than two decades since its publication, Gatrell's *The Hanging Tree* energised the study of the history of crime and punishment in Britain, opening new avenues of inquiry in which scholars including Simon Devereaux, Randall McGowen and Peter King have published excellent work.[23] However, where Gatrell's investigation ended—at the gallows—ours begins. Punishment and spectacle in no way concluded with the death of the condemned, and continuing the journey into the spaces and practices of post-mortem punishment is critical to understanding the history of crime and punishment and also the social history of this period in Britain.

We also position our intervention as a complication, if not a challenge, to the civilising narrative mentioned above. In addition to Elias's civilising progress, Michael Ignatieff's influential *A Just Measure of Pain* argued that from the eighteenth century public and bodily punishments in England were replaced with the moral management of the prison and the prisoner.[24] Not just the formalised, legislated use of post-mortem punishment under the Murder Act, but also the enthusiasm of the crowd who were entertained, enlightened, or affected by the displays of such punishments throughout the life of the Act pushes back against Ignatieff and Elias's frameworks. This should disrupt assumptions of progressive social progress in Britain or metanarratives relying on a civilising trajectory. The criminal corpse always resists being closed down into a single narrative: its meanings are multiple and mutable.

We should note, finally, that our approach pursues many themes which will be familiar to historical sociologists: for example, power, performance, identity, the body, alterity. These themes have been explored widely by many scholars across the humanities and we hope that we have also been able to contribute to those ongoing interdisciplinary conversations.

BODY AND POWER

At the centre of this enquiry is the physical body of the condemned criminal in the long eighteenth century: a body whose owner was tried, found guilty, executed and which was then subject to further punishment by being either 'anatomised and dissected' or 'hung in chains'. Both punishments involved public display and usually resulted in the obliteration of the corpse. However, these grisly post-mortem fates were not reserved exclusively for murderers. For judges, hanging in chains fell within their sentencing repertoire. This punishment was used before the Murder Act, though usually reserved for those condemned for particularly serious crimes (see Chapter 6). The corpses of executed criminals could also make their way into the hands of anatomists and surgeons, but their use did not include public display of the cut body before transfer to the rooms of the surgeons or anatomists as occurred under the Murder Act. Condemned criminals could sell their body by private agreement with the surgeons in exchange for money for themselves or their families. Some criminal corpses were given over to the medical men by the Crown as part of special annual grants (see Chapter 5). Whether mandated by law or handed down at the discretion of a judge, harming the corpse of an individual executed by the state was intended to increase the horror and deterrence of the punishment. This book examines attitudes to the post-mortem punishment of the criminal body as a prism through which beliefs about the human body and its death are refracted.

So, why should the threat of dissection or hanging in chains be considered sufficiently effective to justify its enshrinement in law in the mid-eighteenth century? Although the number of people punished under the Murder Act was fairly small, the social and cultural impact of their punishment was high. It has been argued that in premodern Britain the frequency and ubiquity of death resulted in a somewhat desensitised population, for whom an encounter with a dead body was an almost mundane experience.[25] In a period where the mean life expectancy was about 40 years, a figure affected heavily by high infant mortality—about a fifth of the population failed to reach the age of ten—compared to British life expectancy of about 80 years today,[26] most adults would have seen the dead bodies of siblings and at least one parent by the age of 20. However, as social historians including Alan McFarlane and Lawrence Stone have demonstrated, neither the emotional impact of bereavement nor the fear of one's own mortality was diminished by familiarity.[27] Death was common indeed, but the impact of a death, especially a death that occurred under heightened dramatic and emotional circumstances such as murder or execution, was still profound.

At the heart of our fundamental question is an important interpretive tension: is the criminal body the same as any other body except for the circumstances in which it finds itself, or is there something inherent in the body which determines its criminality and makes the criminal body an essentially different thing to a noncriminal one?

During the period of the Murder Act there were two schools of thought on that issue: the environmental and the anthropological. It was not until the early nineteenth century that there arose a coherent 'scientific' discourse on the determination of character by somatic features, considering criminality to be an essential variable of personhood. In the middle and later nineteenth century, racialised discourses of phrenology and anthropometry were dominant in criminology. Shane McCorristine studied the collections of skulls assembled by both phrenologists and their opponents, in order to demonstrate or refute the anatomical origin of criminality and other personality traits. Of the collection established by Francis Gall himself, the founder of phrenology, McCorristine says:

> Gall's own collection of skulls and casts, now mostly at the Musée de L'Homme, Paris, contained specimens of over one hundred criminals. The catalogue listing gives us an idea of what they were intended to demonstrate:

Skull 5600-4-2-3. A soldier who was executed for having killed a prison guard. The organisation which produces proud, unmanageable personalities who cannot bear authority, is very noticeable here.

Cast 5624-35-3-8. A young Prussian boy of fifteen who had an irrepressible tendency toward stealing. He died in a reformatory where he was to spend all his life. Seen before last condemnation by Gall, who considered him incurable.

...Phrenologists constructed hierarchies of cranial types and the criminal was an important piece in the jigsaw of their unscientific and prejudiced system.[28]

According to this view, there was very clearly a criminal body preceding any criminal act. That is to say, one did not become a criminal because one had committed a crime, but one was likely to commit a crime because one had, congenitally, a criminal body. The unfortunate man with a low forehead, small eyes, and a sloping profile was pretty much doomed by birth to be a criminal. Whole races were judged to have criminality literally in the blood. Such a view supported imperialist and expansionist ideologies of racial hierarchy since, unsurprisingly, the northern European racial type was (supposedly) endowed with superior intelligence, character and a proclivity for civilisation. Further down the scale came Mediterranean, Semitic and Asian types, and at the bottom the 'negro' and 'aboriginal' races. This classification was buttressed by huge collections of bodily measurements and indeed, huge comparative collections of skeletal examples. Scientists and anthropologists created these collections through grave robbing on nearly every continent—clandestinely or coercively obtaining human remains that were 'interred as people and ... extracted as resources'.[29] People made vulnerable through colonial dispossession and enslavement were particularly frequent targets, and the power dynamics exploited to build and interpret these collections inheres in them today. As Megan J. Highet has identified, there are strong parallels between 'collecting' human remains for these purposes and obtaining the corpses of the unwilling or unknowing for medical practice and research.[30]

Although the scientific legitimation of racism through biological anthropology did not become dominant until the mid-nineteenth century, the precursors to this movement were established earlier in the period. Phrenology, for example, was already popular by the 1820s, and appeared to offer a scientific and rigorous approach to determining character. By the 1830s it had become normal to make plaster models of the heads of executed criminals for the purposes of phrenological study. The case of Eugene Aram is an instructive one.[31] Aram's actual skull was, in the 1830s, the focus of a phrenological investigation into whether he was, in fact, guilty of the murder for which he was executed in 1759 or whether he was rather a gentle, scholarly type and therefore not, *by reason of biology*, capable of killing another person. The existence of another report interpreting Aram's skull as that of a morally weak and venial man, however, was used by detractors of the 'science' to ridicule its methods and claims. The biological determinism of phrenology was not uncontested.

Another approach to criminality in the early nineteenth century was the environmentalist school. Rather than locating criminality in original sin or rooted in a bad bloodline and therefore inherent in a criminal body, environmentalists ascribed criminal behaviour to extrinsic factors.[32] These particularly included the nature of housing and neighbourhood amenities, the influence of family and friends, and the quality of education. The environmentalist approach to crime was implicit in nineteenth-century prison reform, which aimed at removing the offender from their bad surroundings and relocating them into the rational environment of the prison, where rules of silence or segregation minimised the influence of other, more intractable villains.[33] Improvement of the living conditions of the poor, and the extension of education reform also arose from nineteenth-century environmentalist perspectives.[34]

There were, then, broadly two schools of thought: criminals were born, or they were made. According to the first, a criminal body was the inherent and congenital origin of criminality. A criminal was born that way, and condemned by their own body. Or, a criminal body was produced only by committing crime: there was nothing in one's physical makeup to predispose a person to criminality. A further complication comes from the sociocultural, and thus mutable, understanding of what constitutes a crime. Sixty years ago, in Britain, seeking or carrying out any kind of abortion was a crime, as was a consensual sexual act between two men. At the same time, 'light physical chastisement', including beating with a slim stick, of a wife by her husband, was considered acceptable by law.[35] Social attitudes have changed, and so has the law: neither abortion nor homosexuality is a crime, and assault, regardless of the relationship of the people involved, is unambiguously a criminal act. With these changes, various sets of criminal bodies ceased to be criminal. Diachronic change in what makes a criminal body can be even more extreme: in our concluding chapter we consider the case of the men condemned and executed for cowardice during the First World War, who are now commemorated and celebrated as heroes and victims following revisionist campaigns to pardon and remember those who were 'shot at dawn'.

If we accept that criminality is not inherent in the body, but rather arises from the actions of the individual in combination with the judgement of law-keepers and the laws put in place in their society, then what is the source of the power of the criminal corpse, and in what ways can it be considered powerful?

First, the criminal corpse was literally powerful as the source of a magical or medical healing energy. The touch of the newly hanged man's hand was a powerful cure, extensively sought well into the nineteenth century. Parts of the criminal body—its bones, blood or dried flesh—formed the basis of several remedies in Britain and around Europe. The dried hand of a hanged man could be used as a 'hand of glory', which had magical properties to thieves and burglars.[36]

Second, the execution scene and the subsequent display of the criminal body being opened or decaying in a gibbet was socially powerful as a symbolic resource that could be recruited to further particular ends, including the creation or maintenance of certain relationships of power and inequality. This kind of political and social power underlies the changes in the law of punishment and how those punishments were carried out. More subtly, this kind of power affected the way that the public viewed and talked about the drama and exhibition of the criminal corpse, not always in the ways that the legal and governmental authorities intended or hoped.

Third, dead bodies of all kinds, but maybe especially the dead bodies of executed criminals were culturally powerful as a resonant signifier of a bad death, or frightening ghoulishness (even today). One of the key findings of our research is that during the long eighteenth century, many people found the public display of dead and decomposing bodies creepy and ghoulish. Methods of gibbeting and features of the gibbet accentuated the unsettling and disturbing aspects of the gibbeted body. Though a dead body, it remained upright and above ground. Its visibility was enhanced by locating the gibbet in a prominent place, and as close as possible to the scene of the crime. Though a dead body, it moved. The gibbet cage was suspended from the gibbet arm using a hook and a short length of chain, so that it would move in the wind, and turn about. Though a dead body, it made a noise. Contemporaries described the eerie sound of the creaking of chains and the cawing of carrion birds. Though a dead body, hanging above the road it seemed to watch people coming past. Letters, diaries, petitions, and common folktales tell of people's reluctance to pass by a gibbet, especially at night.

Last October, a local supermarket devoted an entire aisle to Halloween paraphernalia. Among the decorative cobwebs and inappropriately horrific children's costumes was a startlingly gruesome hanging gibbet ornament. A semi-skeletal figure dressed in prisoners stripes gripped the bars of a cage and, when the contraption was switched on, croaked 'Let me out', accompanied by some scene-setting rattling chains. This gibbeted figure was a grotesque and mildly frightening piece of Halloween tat, not a pedantically correct historical reconstruction; it was intended to be 'good fun', insofar as murder, execution and humiliation, and post-mortem violations of the body count as fun. But in another way, it would be wrong to draw too sharp a distinction between the past and the present. If the customers of a budget supermarket in the British East Midlands find the gibbeted criminal creepy, rather than an awe-inspiring demonstration of the power of the State and the implacability of Justice, so too did their eighteenth-century forebears. The government that passed the 1752 Murder Act hoped that it would deter criminals by graphically demonstrating the consequences of crime. Much like a farmer nailing up the corpses of shot crows on a field gate, hanging in chains was intended to impress a specific message on the hearts and minds of others. Even at the time, however, creepy nastiness rather than moral reflection was often the result. In this light, the appropriation of a misremembered and 'gored-up' version of the eighteenth-century gibbet for the expanding commercial blood-fest that is twenty-first century Halloween is a fitting tribute to a post-mortem punishment that never quite achieved what its legislators hoped it would.

So the criminal corpse was and is a polyvalent object, capable of being co-opted into subversive discourses. Its meaning was hard to control and it could easily slip from an object of terror to one of pity, or from demonstrating the might of the law to religious sacrifice or folk hero.

The dead body is an especially potent symbol. It is always already freighted with cultural meaning—in some ways a 'hypersignifier'. From the grinning plastic skeletons of Halloween to the ubiquitous artistic and devotional representations of Christ on the cross, the human body does not lose significance when it loses life. Cultural historian Thomas

Laqueur recently compiled a history of the cultural work done by the dead, both as abstracted memories and as surviving bodies or body parts. The history of the work of the dead is, claims Laqueur, a history of how the living invest 'the dead body with meaning and is thus the greatest possible history of the imagination.'[37] He cites historian Richard Cobb who noted 'The most dangerous person at a funeral is the body in the coffin'.[38]

What is true of the dead body in general is especially true of the executed body. That the criminal body is the central focus and principal player in the drama of the execution is well known. What is less well documented, and what this book explores, is how the criminal body continued to play a focal role in the ongoing performance of post-mortem corporal punishment. The signifying power of the corpse is enhanced in particular by two things.

First: Every death has the power to evoke other dead bodies. Each new corpse partakes of a cultural tradition of representing death and the dead. In Christian contexts the most potent of these is the crucified Christ. The next chapter considers how, in the medieval period, the resemblance between the executed criminal and the body of Christ could be enhanced to promote particular interpretations of the execution event, and how an unintended evocation of the Christian sacrifice could undermine other, authoritative, readings.

There is a tension between the unique story of each individual criminal whose body ends up being executed, and the universal body or representative criminal that comes to stand for something more than itself. This is clearly evident in the physical exploitation of executed criminal bodies in the demonstration and development of modern medical science. Practical anatomy and the value of dissections depend precisely on the universality and interchangeability of human beings. A surgeon or doctor must be able to assume that the interior configuration of bodily organs and systems should be predictable and should not vary significantly between people. Similarly the efficacy of ostentatious bodily punishment is wholly consequent upon the representativeness of a single criminal. The identity of an individual as a murderer must exceed their identity as, say, Mary Ann Higgins or John Holloway in order to function as a demonstration of the consequences of crime. They could be any murderer, and any murderer's body could do duty for theirs.

So, each newly executed and dissected or gibbeted body was reminiscent of others that had gone before. Each death carried the memory of earlier deaths; each criminal corpse evoked other criminal corpses; each pained, humiliated and ultimately extinguished person on the scaffold or the dissection table or swaying in a gibbet high above the ground called to mind others that had been witnessed or whose representations were familiar from pictures and stories.

At the same time, the distinct histories and individual bodies of executed criminals were eagerly consumed by the readers of pamphlet literature—the original 'true crime' genre—and the names of particular murderers came to be fossilised in place names and ballads. Sometimes the distance between an actual suffering body and the well-known representations of other executed bodies were what struck the observer: the difference between an ugly corpse with a black, swollen tongue and a smell of urine, and the beautiful depictions of the crucified Christ in glory, or the elegantly composed and artfully lit painting of an anatomy by Rembrandt.

Second: Being dead, the body has limited potential to challenge its co-option into other stories. We have established that the corpse still has power—indeed that is the fundamental argument of this whole book—but its power is inarticulate, inchoate, and requires cultural interpretation and trammelling to shape it to particular ends. Zoë Crossland has unpicked the common trope of forensic study of the dead body, in particular of fictional representations of such study in TV dramas and popular novels.[39] She notes that the metaphors of reading the body and of hearing the dead body speak are popular interpretations of the science of forensic pathology. However, dead bodies do not give clear and unambiguous testimony. They do not 'tell' their stories in any unmediated way, nor are they amenable to being 'read' in any straightforward way. They require interpretation. The forensic pathologist does not simply decode the body.

Similarly, in this book, the corpse does not tell its own story. We do. This has required research, contemplation, conversation, and engaging with the challenges of speaking with and for these histories. And as discussed in the final chapter, we have struggled with the ethical implications of focusing on criminal corpses, not the stories and legacies of victims. Here the corpse offers a number of potentials and constraints, but the dead body itself has limited capacity to resist the narratives and arguments into which it is brought. A dead body does not answer back, but we cannot escape the power of the criminal corpse.

NOTES

1. Reported by the *International Organisation for Migration*, 28 August 2015, 'Mediterranean Migrant Arrivals, Deaths at Sea Soar'. Available at https://www.iom.int/news/mediterranean-migrant-arrivals-deaths-sea-soar (Accessed 1 June 2017).
2. http://www.bbc.co.uk/news/av/uk-politics-33714282/david-cameron-swarm-of-migrants-crossing-mediterranean.
3. A poll a few days after the event found that Britons who had seen the photographs were nearly twice as likely to want to take in more refugees that those who had not (44% of those who had as opposed to 24% of those who had not). The poll was undertaken by BBC Newsnight and reported by ComRes (Communicate Research Ltd.), available at http://comres.co.uk/polls/bbc-newsnight-refugee-poll/ (Accessed 1 June 2017).
4. See, Kristeva, J. (1982), *Powers of Horror: An Essay on Abjection* (New York: Columbia University Press).
5. See, Hodder, I. (1985), 'Postprocessual Archaeology', *Advances in Archaeological Method and Theory*, Vol. 8, 1–26.
6. Amongst others, see Hodder, I. (1986), *Reading the Past: Current Approaches to Interpretation in Archaeology* (Cambridge: Cambridge University Press); Hodder, I. (1989), 'This Is Not an Article About Material Culture as Text', *Journal of Anthropological Archaeology*, Vol. 8, Issue 3, 250–269; Hodder, I. (1992), *Theory and Practice in Archaeology* (London: Routledge).
7. See for example, Sofaer, J.R. (2006), *The Body as Material Culture: A Theoretical Osteoarchaeology* (Cambridge: Cambridge University Press).
8. See for example, Tarlow, S. (2011), *Ritual, Belief and the Dead in Early Modern Britain and Ireland* (Cambridge: Cambridge University Press); Cherryson, A., Crossland, Z., and Tarlow, S. (2012), *A Fine and Private Place: The Archaeology of Death and Burial in Post-Medieval Britain and Ireland* (Leicester: University of Leicester Press).
9. Tarlow, *Ritual, Belief and the Dead*, pp. 145–152.
10. 25 Geo II c.37. An Act for Better Preventing the Horrid Crime of Murder.
11. The replacement of murder by poverty as the 'crime' most likely to result in dissection is an intriguing aspect of political history with some contemporary resonances. On this subject we recommend the excellent work of Richardson, R. (2000), *Death, Dissection and the Destitute* (Chicago: University of Chicago Press, 2nd Edition); Hurren, E.T. (2012), *Dying for Victorian Medicine: English Anatomy and Its Trade in the Dead Poor, 1832–1929* (Basingstoke: Palgrave Macmillan).

12. For published work from this strand see, Ward, R. ed. (2015), *A Global History of Execution and the Criminal Corpse* (Palgrave Macmillan), available at http://www.palgrave.com/us/book/9781137443991; Ward, R. (2014), *Print Culture, Crime and Justice in Eighteenth-Century London* (London: Bloomsbury), available at http://www.bloomsbury.com/uk/print-culture-crime-and-justice-in-18th-century-london-9781472507112/; King, P. ed. (2016), *Punishing the Criminal Corpse, 1700–1840* (Palgrave Macmillan), available at http://www.palgrave.com/de/book/9781137513601; King, P. and Ward, R. (2015), 'Rethinking the Bloody Code in Eighteenth-Century Britain: Capital Punishment at the Centre and on the Periphery', *Past and Present*, Vol. 228, 159–205; Ward, R. (2015), 'The Criminal Corpse, Anatomists and the Criminal Law: Parliamentary Attempts to Extend the Dissection of Offenders in Late Eighteenth-Century England', *Journal of British Studies*, Vol. 54, 63–87; Ward, R. (2012), 'Print Culture, Moral Panic, and the Administration of the Law: The London Crime Wave of 1744', *Crime, History & Societies*, Vol. 16, 5–24.

13. For published work on this strand, see Hurren, E.T. (2016), *Dissecting the Criminal Corpse: Staging Post-execution Punishment in Early Modern England* (Palgrave Macmillan), available at http://www.palgrave.com/br/book/9781137582485; Hurren, E.T. (2013), 'The Dangerous Dead: Dissecting the Criminal Corpse', *The Lancet*, Vol. 382, No. 9889.

14. For published work on this strand see, Tarlow, S. and Dyndor, Z. (2015), 'The Landscape of the Gibbet', *Landscape History*, Vol. 36, Issue 1, 71–88; Tarlow, S. (2017), *The Golden and Ghoulish Age of the Gibbet in Britain* (Palgrave Macmillan), available at http://www.palgrave.com/de/book/9781137600882.

15. See, Davies, O. and Matteoni, F. (2017), *Executing Magic in the Modern Era: Criminal Bodies and the Gallows in Popular Medicine* (Palgrave Macmillan), available at http://www.palgrave.com/de/book/9783319595184; Davies, O. and Matteoni, F. (2015), '"A Virtue Beyond All Medicine": The Hanged Man's Hand, Gallows Tradition and Healing in Eighteenth- and Nineteenth-Century England', *Social History of Medicine*, Vol. 28, Issue 4, 686–705.

16. For published work on this strand see, McCorristine, S. (2017), *Interdisciplinary Perspectives on Mortality and Its Timings* (Palgrave Macmillan), available at http://www.palgrave.com/de/book/9781137583277; Matteoni, F. (2016), 'The Criminal Corpse in Pieces', *Mortality*, Vol. 21, Issue 3, 198–209.

17. See, Tomasini, F. (2017), *Remembering and Disremembering the Dead* (Palgrave Macmillan), available at http://www.palgrave.com/de/book/9781137538277.

18. See, McCorristine, S. (2014), *William Corder and the Red Barn Murder* (Palgrave Macmillan).
19. Amongst others, see, O'Gorman, F. (2016), *The Long Eighteenth Century: British Political and Social History 1688–1832* (London: Bloomsbury, 2nd Edition). The long eighteenth century is the most popular of an extensive suite of long centuries in historical scholarship from the second up to the twentieth. There is also a sprinkling of short centuries, but to our knowledge no fat or thin ones.
20. Norbert Elias, *The Civilizing Process* (first published in German in 1939 as Über den Prozeß der Zivilisation). It was republished in English translation in 1969 after which it gained widespread prominence.
21. See for example, Tarlow, S. (2017), *The Golden and Ghoulish Age of the Gibbet in Britain* (Palgrave Macmillan); Hurren, E.T. (2016), *Dissecting the Criminal Corpse: Staging Post-execution Punishment in Early Modern England* (Palgrave Macmillan).
22. See, Gatrell, V.A.C. (1994), *The Hanging Tree: Execution and the English People 1770–1868* (Oxford: Oxford University Press).
23. See for example, McGowen, R. (2004), 'The Problem of Punishment in Eighteenth-Century England', in Devereaux, S. and Griffiths, P. (eds.), *Penal Practice and Culture 1500–1900: Punishing the English* (Basingstoke: Palgrave Macmillan); Devereaux, S. (2009), 'Recasting the Theatre of Execution: The Abolition of the Tyburn Ritual', *Past & Present*, Vol. 202, Issue 1, 127–174; King, P. (2006), *Crime and Law in England 1750–1850. Remaking Justice from the Margins* (Cambridge: Cambridge University Press).
24. See, Ignatieff, M. (1978), *A Just Measure of Pain: The Penitentiary in the Industrial Revolution, 1750–1850* (New York: Pantheon Books).
25. See the discussion in Linton, A. (2008), *Poetry and Parental Bereavement in Early Modern Lutheran Germany* (Oxford: Oxford University Press), pp. 1–5.
26. Statistics on changes in life expectancy are available from the Office for National Statistics (ONS) and are summarised here, http://visual.ons.gov.uk/how-has-life-expectancy-changed-over-time/ (Accessed 1 June 2017).
27. See, MacFarlane, A. (1986), *Marriage and Love in England: Modes of Reproduction 1300–1840* (Oxford: Blackwell); Stone, L. (1977), *The Family, Sex and Marriage in England 1500–1800* (London: Penguin Books).
28. See the Power of the Criminal Corpse blogpost, 'Being Disturbingly Informative', written by project member Shane McCorristine on the work of phrenologists and the continuing fate of collections of human skulls and casts. Available at http://staffblogs.le.ac.uk/crimcorpse/2016/10/31/disturbingly-informative/ (Accessed 1 June 2017).

29. See, Highet, M.J. (2005), 'Body Snatching & Grave Robbing: Bodies for Science', *History and Anthropology*, Vol. 16, Issue 4, 415–440.
30. Ibid.
31. On the case of Eugene Aram, see Tarlow, S. (2017), *The Golden and Ghoulish Age of the Gibbet in Britain* (Palgrave Macmillan); Tarlow, S. (2016), 'Curious afterlives: The Enduring Appeal of the Criminal Corpse', *Mortality*, Vol. 21, Issue 3, 210–228. Aram is also discussed in Chapter 8 of this book.
32. See for example, Driver, F. (1988), 'Moral Geographies: Social Science and the Urban Environment in Mid-Nineteenth Century England', *Transactions of the Institute of British Geographers*, Vol. 13, Issue 3, 275–287.
33. See, Ignatieff, M. (1978), *A Just Measure of Pain: The Penitentiary in the Industrial Revolution, 1750–1850* (New York: Pantheon Books).
34. See, Tarlow, S. (2007), *The Archaeology of Improvement in Britain 1750–1850* (Cambridge: Cambridge University Press).
35. See, Siegel, R.B. (1996), '"The Rule of Love": Wife Beating as Prerogative and Privacy', *The Yale Law Journal*, Vol. 105, Issue 8, 2117–2207.
36. See, Davies, O. and Matteoni, F. (2015), '"A Virtue Beyond All Medicine": The Hanged Man's Hand, Gallows Tradition and Healing in Eighteenth- and Nineteenth-Century England', *Social History of Medicine*, Vol. 28, Issue 4, 686–705.
37. See, Laqueur, T.W. (2015), *The Work of the Dead: A Cultural History of Mortal Remains* (Oxford: Princeton University Press), quote at p. 17.
38. Ibid., p. 18.
39. See, Crossland, Z. (2009), 'Of Clues and Signs: the Dead Body and Its Evidential Traces', *American Anthropologist*, Vol. 111, Issue 1, 69–80.

CHAPTER 2

The Power of the Criminal Corpse in the Medieval World

GETTING MEDIEVAL ON YOUR ASS

It is ironic that one of the most notoriously violent characters in one of the most notoriously violent films of the late twentieth century, Marcellus in Quentin Tarantino's *Pulp Fiction*, is remembered for introducing his most gruesome brutality with a warning that he was about to 'git medieval on your ass'.[1] Violent punishment seems to have been both more common and bloodier in Tarantinoland than in actual medieval Europe, when justice was often more likely to take the form of a fine or be put in the hands of God than anything involving ironmongery or slow torture. But the assumption that medieval punishment was bloodthirsty and was spendthrift of lives and limbs was not limited to the world of popular film. Even academics, outside the tradition of medieval specialists, sometimes employ this stereotype.

The assumed brutality of medieval justice was given a fillip by the well-known and widely cited model of the history of punishment in the West developed by Michel Foucault. Foucault's *Discipline and Punish* famously opens by contrasting the extreme pain and cruelty of the execution of the regicide Damiens in 1757, with the official documentation of a reforming prison discipline dated only eighty years later.[2] Foucault's thesis is that 'modern' punishment is aimed at reforming criminals into conforming members of capitalist society through the enforcement of authoritarian disciplinary regimes. This, he contrasts with an

© The Author(s) 2018   29
S. Tarlow and E. Battell Lowman, *Harnessing the Power of the Criminal Corpse*, Palgrave Historical Studies in the Criminal Corpse and its Afterlife, https://doi.org/10.1007/978-3-319-77908-9_2

earlier 'medieval' kind of bodily punishment which was retributive and deterrent, and which exercised State power in an overt and demonstrative show of force. His model of changes to criminal punishment has been massively influential (less so among historians than in many other areas of the social sciences and humanities) but, as Dean notes, the assumption that a piece of mid-eighteenth-century penal theatre represents the direct and unchanged legacy of the Middle Ages is wrong; the 'Foucault effect' perpetuated a number of misconceptions about medieval justice.[3] For a start, although one can easily find examples of extreme bodily cruelty in punishment, one of the striking features of medieval law in action is its reliance on fines and even imprisonment. In the early medieval period, and up until about the twelfth century, many crimes, even serious and violent ones, were amendable through the payment of compensatory 'wer' or 'wergeld'—literally 'man-money'—to the victim or their kin.[4] Reynolds notes that of the 178 lawsuits of tenth to eleventh century date considered by Wormald, only six mention capital punishment; the majority of crimes up to and including murder, were punished with fines.[5] Medieval punishment, therefore, should not be seen as a poorly differentiated 'premodern' state of culture, of which early modern spectacular justice was a manifestation.

The same is true of medieval beliefs about death generally. The work of another influential French thinker—also not a historian—is responsible for perpetuating the view that 'medieval' death was part of an organic, undifferentiated, premodern mindset. Phillippe Ariès claims that death in the Middle Ages was the same as death for 'the ancients' and probably in prehistory too.[6] It also, says Ariès, characterises the Russian peasants described by Tolstoy as calmly accepting their own death without fear or resistance, and some other naïve and uneducated people in modern history. Ariès's 'tamed death' is a death that is expected, not feared and not agonised over. It is a simple, almost animal, acceptance of the inevitable. Ariès's view is problematic on a number of levels. First, he offers not a shred of evidence that such an attitude characterises his homogenous 'prehistory', and evidence for the Middle Ages is anecdotal and promiscuous in time and place, with a concentration on literary sources. Ultimately, Ariès's medieval functions mostly as a foil for the development of cultural attitudes during modernity. Accordingly, the 'premodern', as for Foucault, is ahistorical and almost outside culture, an undifferentiated mass of hessian-wearing, mud-bespattered peasantry persisting down the ages.

In fact, attitudes to death in the Middle Ages, like attitudes to punishment at that time, are not reducible to any unified and coherent position that would be recognisable throughout the period and throughout Europe, let alone beyond it.

Early Medieval Death and the Context of Punitive Death

The Middle Ages (a term used in this chapter synonymously with 'the medieval period') are customarily divided into early and late, or early, high and late periods. In England, the early Middle Ages embrace the years between the end of Roman rule in the fifth century AD and the eleventh-century Norman conquest, whilst the late period lasts until about the time of the Reformation in the sixteenth century. The early period can be further divided into pre-Christian and Christian times. Historical sources on the history of the body, death and criminal execution are scanty for this period, but the shortfall of historical evidence of medieval criminal bodies is to some extent made good by a wealth of interesting and provocative archaeological evidence, particularly in the Anglo-Saxon areas of south and east England.

Where historians can start from the fact that the body of a criminal known from historical records must have been disposed of somehow, archaeologists, especially in earlier periods, start with the disposed body and work backwards to suggest that it might be the body of a criminal. In such cases the inference of criminality is mostly made when a body has been subject to non-normative mortuary treatment, conventionally known by archaeologists as 'deviant burial'.[7]

Deviant burial in the Anglo-Saxon parts of Britain during the early medieval period has been the subject of extensive research recently.[8] During the earliest part of the Anglo-Saxon period the victims of judicial execution are hard to recognise archaeologically due to the diversity of normal burial practices. However, in the post-conversion period 'execution cemeteries' are clearly identifiable, characterised by regularly occurring non-normative practices including prone burial (in which the body is laid flat and facing down), multiple interments, decapitation, evidence of restraint (tied wrists and ankles), shallow and cramped burial and ante- and peri-mortem mutilation (i.e., damage to the body occurring before or around the time of death). Execution cemeteries frequently contain burials of varied orientation, often intercutting one another. Intercutting burial is evidence that the locations of previous burials were not remembered, marked out or consciously avoided afterwards, in contrast to community cemeteries which generally buried all people with heads to the west, supine and in neat rows. Execution cemeteries are frequently sited on or near boundaries and close to older or contemporary earthworks.

Reynolds infers from these burial practices a clear distinction between 'members' of the community and 'others'; otherness being signalled by prone burial and decapitation.[9] Some early medievalists suggest that these practices, and others such as 'weighing down' the corpse with stones, might also have been attempts to prevent the dead from returning to trouble the living.[10] His hand list of 27 execution cemeteries shows their frequent occurrence in marginal locations, another clear statement of sociocultural liminality.

Overall, the power of the State is increasingly evident from the seventh century, but there is also evidence, in the persistence of local customs of burial and stigmatisation, of continuity from well-established local traditions.[11]

Foxhall Forbes puts this evidence into a religious context, and demonstrates how, in the Christian Anglo-Saxon period, religious belief shaped and was shaped by popular understandings and practices as well as recondite theological disquisitions.[12] The tradition of burying people with their heads to the west, for example, is pretty much continuous from the Neolithic or Bronze Age through to the medieval period and indeed to the present day. Although sometimes glossed as the correct orientation for a Christian resurrection, the custom was already ubiquitous thousands of years before Christianity.

Late Medieval Death and the Changing Context of Punishment

Over the late medieval period, the structures of the Church became ever more elaborate and more aspects of private life and practice came to be controlled by the Church and by secular law, including bodily processes such as consumption and sexuality.[13] Thus, religious laws specified periods of fasting and complex dietary restrictions; codes of celibacy and controlled sexuality were specified for different orders, genders and times. However, the manner of death and burial was subject to a lesser degree of formal control, and the ideal or stereotyped normal death seems to have changed little over the whole medieval period. Around the twelfth century, however, approaches to crime and punishment altered. The shift in justice was from an oppositional to an inquisitorial framework. Whereas in the earlier period an accusation would be adjudicated by God through an ordeal, in the later Middle Ages trials came to be about reviewing evidence and making a judgement. At the same time, a new code of punishment emerged. Serious crimes could no longer be compensated by the payment of a fine.

If a jury convicted a person of a serious crime, their judgement took the place of the corporeal ordeal, and punishment was then handed down and carried out. Punishment options included imprisonment, payment of fines or forfeiture of estate, and various corporal sanctions including whipping, stocks, pillory, branding or the removal of a body part such as a hand or foot, or capital punishment, normally by hanging, though certain crimes were punished by burning.[14] As discussed below, the capital punishment that followed a conviction for treason was subject to special symbolic elaboration.

Medieval Bodies: Living, Lived, Dead and Damned

Harris and Robb note that scholarship on 'the medieval body' is fragmented—perhaps more so than the history of the body in any other time period.[15] They identify three kinds of 'medieval body', across which a number of other themes cut. The three kinds are the theological, the scientific and the actual lived body. Cross-cutting themes include gender, normal and abnormal bodies (monsters and so on), and metaphor—both the metaphors by which the body is described and body metaphors as used to describe other things such as the organisation of the cosmos or the political system. Broadly, scholarship about the medieval body tends to focus on one kind of body, and/or one theme, though much interesting thought has emerged from exploring the tensions between different and often incompatible beliefs about the body.[16]

The question 'what did medieval people believe about the body?' is, unsurprisingly, impossible to answer. Not only does the label 'medieval' encompass more than a thousand years of history over three continents, but it is also fair to assume that the preoccupations of a Merovingian peasant woman, a fourteenth-century bishop and a twelfth-century Irish poet were necessarily very different. Moreover, the body was significant in context rather than as an encompassing abstraction. It is unlikely that the question 'What do you believe about the body?' would have made any more sense to a medieval person than it would to most non-academics today.

In this chapter, medieval beliefs about the lived body, that is, the body needing medical attention, or giving birth, or eating, drinking, copulating, excreting, fighting, crafting or riding, are not our main concern. Nor is it highly relevant to look at the gendered body, or its age categories, or at animal bodies. Rather, our focus here is particularly the body in death, and more particularly, the body whose death is the result of having committed a criminal act or being subject to the processes of law. As we shall see, in the Middle Ages the criminal body and the operation of justice were inseparable from religious beliefs about sin and judgement. One of the main questions addressed in this book is how the various contextual discourses in which the criminal corpse features—religion, science, magic, social order, political power and so on—relate to each other. We suggested in the previous chapter that in the medieval period those discourses often mapped very closely onto each other, and that, although context would have affected the kind of belief discourse that was prevalent, categories that became very different later on were not necessarily distinguished during the Middle Ages. These included religion and magic, for example, or State and divine ordering.

DEATH AND THE DEAD BODY IN THE MEDIEVAL WORLD

There is, then, no single or unified 'medieval belief about the body'. Different bodies are relevant to different kinds of discourses, at different times and places. And just as modern scholarship on the Middle Ages is fragmented by discipline, tradition, and approach, so in the medieval period there were also disagreements and variations. There were, however, broad areas of shared 'background consensus... embodied in shared terms of disagreement'.[17] These areas of consensus and overlap were greater during the Middle Ages than later on and constituted the kind of necessary commensurability that made disagreement possible. Among these shared taken-for-granteds was a dualistic and oppositional belief in body and soul as a cosmological organising principle. Where the body was temporary, sinful and earthly, the soul was eternal, unsullied and heavenly. In modernity a further dichotomy emerged, aligning on one side the body, the heavy and the material, and on the other the soul, the immaterial and insubstantial. In the Middle Ages, however, there is little doubt that the soul or spirit was no less solid and material than the body. There is a clear tradition of regarding the body, when opposed to the soul, as insignificant at best, and a vile, polluting source of sin at worst.

This kind of somatophobia, related to a profoundly misogynist philosophical outlook, reached its apogee in the early modern period but is built on the opposition between, and differential valuation of, body and soul that had had its roots in the medieval period, and indeed earlier.[18]

At death, the soul departed from the body (though as we shall see, this separation was sometimes incomplete and occasionally reversible). In medieval art, this departure is conventionally depicted as a naked child rising from the body at the moment of death, and being taking up by angels (Fig. 2.1). The dead body was a body without a soul, and was thus a thing to be despised. An early fourteenth-century Franciscan preacher said that 'nothing is more abhorrent than a corpse'.[19] Without the presence of the divine, a body was just an object. Because of its polluting nature, he continued, a dead body could not be put into water or hung in the air where it would spread contagion, but needed to be buried in the earth, and the ground tamped down well 'so that it may not rise again'.[20]

However, medieval beliefs about the body were not consistent or unambiguous. A parallel tradition suggests that the dead body retained some kind of what Horrox calls 'awareness' after death.[21] Katherine Park distinguishes between a northern European recognition that the new corpse retained some 'life-force' until the flesh decayed, and an Italian position that held that life was extinguished utterly with the final breath.[22] The care taken to prepare and place the body in medieval Europe might be evidence of this. In the case of members of highborn families with financial resources this might mean that the body was divided after death so that its parts could be put to rest in more than one location, reflecting the emotional attachments of the individual who had died. The resulting traditions of 'heart burial', known as *mos teutonicus* where an embalmed heart was taken to another location than the rest of its body, was considered repugnant by Italian Pope Boniface VIII who banned the practice in 1300.

Not only the body but also the late medieval soul was consistently described in bodily terms; it was, to use Bynum's word, 'somatomorphic'.[23] From the representation of the departing soul as a small body that comes out of a person's mouth at the moment of death, to the experiences of the soul as it journeys through the geography of the afterlife, the experience of the medieval soul is essentially a corporeal one. The separated soul, for example, the soul after death as it progressed to the afterlife, experienced bodily discomforts and confronted physical obstacles, such as thorny moors, rivers of water and of fire.[24]

Fig. 2.1 A man dies and his soul ascends to heaven. Etching by Karel van Mallery after Jan van der Straet (Wellcome Collection)

Moreover, sometimes experience and identity post-mortem was directly attributed to the same body as the earthly one now lying in the ground. Although according to theological thinking the body in the ground should be empty of personal meaning and spiritual significance, a number of practices suggest that it retained considerable identity and importance. For example, the practice of partitive or heart burial, where the entrails, heart and sometimes head were buried separately from the rest of the body for emotional rather than pragmatic reasons, is evidence that the corpse was still thought to affect and be affected by its placement and treatment.[25] Similarly, the veneration of saintly relics, well studied by Patrick Geary and recently reviewed by Walsham, demonstrates that spiritual and personal 'essence' inheres in the body as well as the separated soul.[26] Both heart/partitive burial and the holy power of saintly relics parallel the somatic kind of spirituality that also informed key medieval religious practices, such as transubstantiation, which depended on the miraculous manifestation of the actual body of Christ.[27]

Westerhof describes how medieval attitudes to the body after death were shaped far more profoundly by religion than our beliefs are today.[28] In the Middle Ages, death was conceived of more as a transition than an end, and therefore it was not death itself but dying in sin that was the really frightening prospect.[29] Accordingly, proper management of that transition, minimising the amount of sin, and thus the time spent atoning for it in Purgatory, and maximising the soul's prospects for resurrection, was of crucial importance.

The ideal death, according to the *ars moriendi* (art of dying) manuals that emerge towards the end of the period, was one that was fully accepted and prepared for (Fig. 2.2).[30] The key preparations were not secular concerns like the disposition of property or funeral arrangements, though these might also be considered, but spiritual ones.[31] Ideally, the death itself takes place peacefully in the heart of family and community, if possible in one's own bed with kin and clergy nearby. This is a death that is predicted, that proceeds slowly—perhaps rather too slowly by modern standards, as the pious final thoughts and prayers can go on for hundreds of pages—giving ample time to prepare the soul, as the organs and powers of the body close down in an orderly and predictable way. A thirteenth-century English verse describes the bodily processes of death thus:

Fig. 2.2 Woodcut illustration from '*Questa operetta tracta dell arte del ben morire cioe in gratia di Dio*' 1503 (Wellcome Collection)

Wanne mine eyhnen misten,
And mine heren sissen,
And my nose coldet,
And my tunge foldet,
And my rude slaket,
And mine lippes blaken,
And my muth grennet,
And my spottel rennet,
And mine her riset,
And mine herte griset,
And mine honden bivien,
And mine fet stivien –
Al to late! al to late!
Wanne the bere is ate gate.

(When my eyes mist/And my hearing hisses/And my nose gets cold/And my tongue folds/And my face slackens/And my lips blacken/And my mouth grins/And my spittle runs/And my hair falls out/And my heart shudders/And my hands shake/And my feet stiffen/All too late! All too late!/When the bier is at the gate).[32]

The execution of a criminal might at first appear to be the very opposite of a good death—a death with crime or sin on one's conscience, violent, away from home and rejected by community. However, as we shall see, medieval judicial execution was not designed to punish the soul in any way and, in fact, could even be seen as a merciful act which would, if anything, improve the malefactor's chances of salvation.

Scary Monsters

Most of all, the materiality of the soul, and the frequent slippage between the earthly and the heavenly body is evident in ghost beliefs. In the medieval period the ghosts and revenant spirits of the dead do not manifest as whispy, translucent, floaty spectres, nor little lights or funny feelings. Rather, as Joynes' extensive anthology of medieval ghost stories demonstrates, the dead are likely to take very solid form—of cadavers, beasts or men, and often with monstrous features.[33] They might violently attack the living or attempt to have sexual relations with them. The ghosts of the dead might also take the form of their dead bodies, especially in English high medieval ghost stories, where the body of the deceased is often the medium of communication between the living and the dead. William of Newburgh's *Historia Rerum Anglicorum* and the fragmentary tales of the fourteenth-century monk of Byland, both contain stories of vexatious ghosts who harass their kin and neighbours until they are laid to rest by digging up the body and placing a scroll of absolution in their grave.[34] In a thirteenth-century German story related by Caesarius of Heisterbach, a living knight tries to protect the ghost of a woman who is being hunted by diabolical figures. He attempts to hold onto the woman, but she struggles free and the knight is left with only a handful of her hair. Since he recognised the woman as a lady who had recently died and was known in her lifetime for unchaste behaviour, he orders her grave to be opened and discovers her body to be missing a clump of hair.[35] The revenant body and the formerly living corporeal body are one and the same.

In most medieval tales, the ghost has a purpose in haunting the living. Commonly, this is to warn a sinful person to mend their ways lest they suffer the same purgatorial pains as the deceased, who now regrets that they did not repent and reform when alive, or to ask the living person to obtain posthumous absolution for sins of the deceased, usually through prayers or masses or by getting a written absolution from a priest. Sometimes the living are asked to rectify a particular wrong as when, in one tale, the ghost of a man appeared to a traveller to ask that his heirs return some sheepskins he had stolen from a widow and a parcel of land that he had obtained by deception. In Purgatory, the ghost had found himself condemned to wear the stolen sheepskins which were burning hot against his skin, and to carry the whole crushing weight of the field on his back.[36]

The majority of ghosts in religious *exempla* and courtly tales were not criminals who had been accorded the justice of the courts, but sinners whose sins had gone undiscovered or unpunished in life. This adds some weight to the suggestion that medieval judicial punishments of the body could act as payment of a debt of atonement that would otherwise be paid in Purgatory.

Magic and Mummia

Because the actual material body was imbued with spiritual power, the material body was also a potent source of curative and totemic magico-religious agency. As Gilchrist has observed, magic and religion in the Middle Ages were not 'mutually exclusive categories', nor were either of them separable from medicine.[37] Most archaeological evidence of magic pertains to the use of magical objects to protect the dead or to mediate the relationship between the living and the deceased. However, the use of the dead body as a place of magical divination was also known, albeit as a sin according to a twelfth-century penitential.[38]

As we shall see in chapter seven, the magical or superstitious use of the criminal corpse does not end with the development of medical science in modernity; medicine and magic continued to overlap well into the nineteenth century and arguably even to the present day.

Crime Is to Sin as Punishment Is to Penance

Criminal justice in the late medieval period, perhaps more than at any other time, was inextricable from religious law. This went beyond an association between Canon (Church) law and Common or customary (state or local) law. Rather, it placed human justice in the same conceptual sphere as divine justice. Crime was an infringement of God's laws as much as of human laws, and therefore secular punishments were not just analogous to holy penance but on a continuum with it.

The late medieval period doctrine of Purgatory introduced an important symbolic territory to the mystical geography of the afterlife.[39] While saints and martyrs had always been able to travel directly to heaven, and unrepentant evil-doers and unbelievers would go directly to eternal torment, what of the majority of people: the not-very-bad? Purgatory provided a temporary stage on the way to redemption: a place where sins could be burned away and bad thoughts paid for. The pains of Purgatory were undeniably horrible, but they were finite and, usefully, of variable duration capable of being affected by the intervention of saints, or reduced by masses and prayers said by the living, and by penances undertaken or indulgences purchased before death.[40] According to some medievalists, pain in the late Middle Ages was a blessing from God, and the means to atonement and redemption.[41] Agony in this world reduced the bill of pain to be paid in the next. Suffering on earth purified and cleansed the sinful soul. This is the principle that underlay the practices of medieval orders of flagellants and other mortifiers of the flesh.

In the case of criminals, a sentence of corporal or capital punishment, especially if it involved intense or prolonged pain, could be not only a punishment but also a spiritually redemptive blessing. A painful and brutal death could, on its own, constitute a pathway to salvation. The story of Engelbert of Cologne, though he was not a criminal, illustrates this. Engelbert was an early thirteenth-century archbishop.

Although he was a man of the Church, he was not a particularly good or virtuous one, and was living a not-very-good, not-very-bad life when in 1225 he was murdered in a bungled abduction plotted by his own cousin, Frederick of Isenberg. He was set upon in a narrow gorge by a gang of armed men while travelling back from consecrating a church and received 47 stab wounds. When his retainers, who had fled the scene, returned to find his dead body, they placed the corpse on a dung cart and brought it to the nearest church, where immediately it began to work healing miracles, restoring the health of those who attended it. Seventy-nine miracles were associated with his relics over the next ten years. The author of Engelbert's vita, Caesarius, says,

> The sanctity which he lacked in life was replenished in full by his death; and if he was less than perfect in his manner of living, he was nonetheless made holy through his suffering.[42]

Engelbert's sanctity then, owed nothing to either his good deeds or his piety. It was entirely the especially gruesome manner of his death that made him holy. His actual material body was transformed into a thing of holiness—his own blood anointed his body in the same places that holy oil would have been used for the last rites attending a more peaceful death.[43]

A year later, Engelbert's murderer, his cousin Frederick, was captured and put to death. He died by breaking on a wheel, penitent, patient, silent and in prayer. Jung notes the symmetry between the two deaths.[44] In Frederick's case, his bodily fragmentation allows the possibility of redemption. The wheel of fortune has turned and the worst of criminals—a man who was responsible for the death of an archbishop and a kinsman—dies in hope of resurrection. The first shall be last and the criminal who dies in pain and shame, like the thief crucified next to Christ, shall be with Him in Paradise.

The redemptive potential of the awful death in the age of the glorification of bodily suffering meant that a criminal execution was an ambiguous spectacle. Its aim was to deter, through public, visible suffering and humiliation, but what the mortified body evoked was also the holy purification of pain. The death of Christ is 'far and away' the most frequently represented death in medieval art, whilst the archetype and the primary association for the late medieval execution crowd was the body of Christ in his passion (Fig. 2.3).[45] Art historian Mitchell Merback notes that late medieval depictions of the passions of Christ owe much to studies

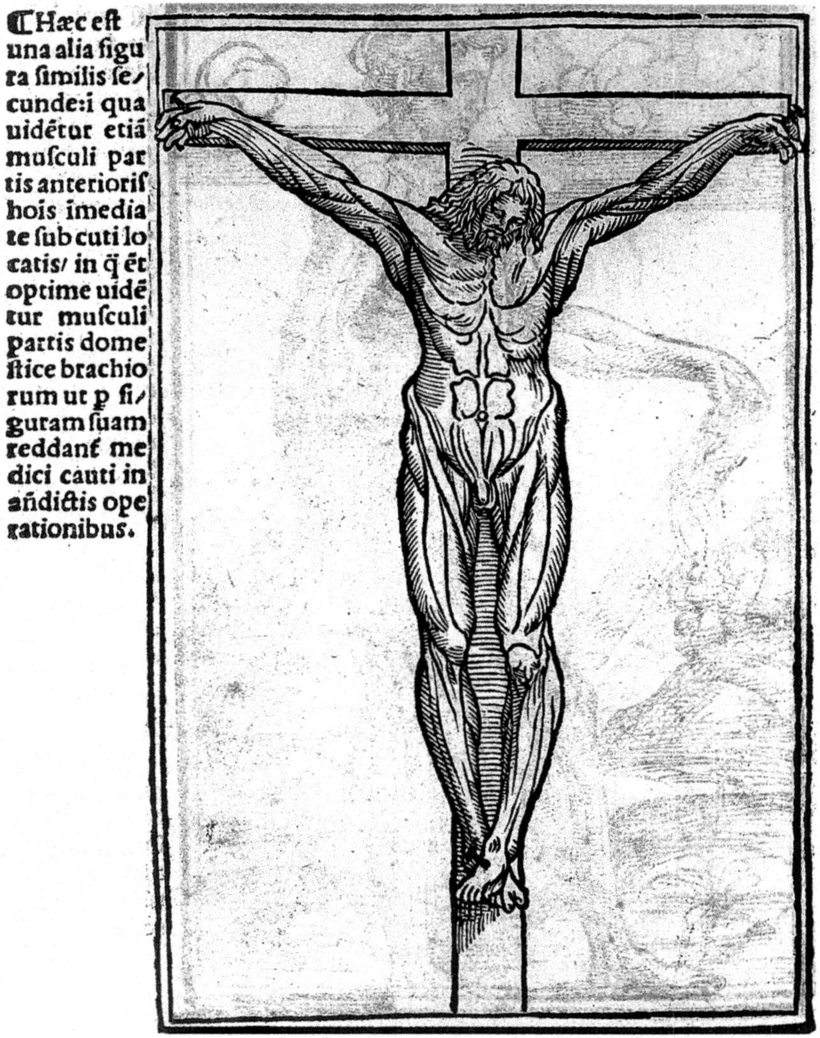

¶Hæc eſt
una alia figu
ra ſimilis ſe/
cunde:i qua
uidétur etiã
muſculi par
tis anterioriſ
hois imedia
te ſub cuti lo
catis/ in q̃ ēt
optime uidẽ
tur muſculi
partis dome
ſtice brachio
rum ut p fi/
gutam ſuam
reddanf me
dici cauti in
añdictis ope
rationibus.

Fig. 2.3 Crucified écorché figure, early sixteenth century (Wellcome Collection)

of criminal bodies hanging or broken (as we will see in Chapter 7, during the time of the Murder Act, the flayed body of a murderer was used as the model for a depiction of the crucified Christ).[46] Meditations on Christ's passion emphasised the bodily aspects of his experience, just as the witnesses to an execution focused on the body of the condemned, 'trembling, sweating, resisting, gesturing, crying, ejaculating blood'.[47] Christ's death, though a criminal execution, was nevertheless a 'good death'—in fact the model of the good death: he 'died a criminal, but he died well' as Binski notes.[48] Other criminal deaths could thus be evaluated according to how close they came to the death of Christ. Did the condemned bear pain with patience, penitence, prayer and hope?

Similarly, dramatic enactments of the crucifixion, the late medieval 'passion plays' which were popular throughout Europe as both pious acts and popular entertainments, emphasised the torture and physical suffering of Jesus, to the point that actors playing Jesus and the thieves were sometimes in danger of their lives.[49] For this reason, all executions had as their ultimate reference point the body of Christ on the cross; and the pain of the condemned was not only an alienating or vengeful outcome of secular justice, but also the basis of an empathetic bond between spectator and sufferer.[50]

As noted before, death by execution was the ultimate known and scheduled death. Death at an appointed moment allowed the subject to repent, to confess, to pray, to prepare their soul as best they can. Execution shared this feature with the ideal, expected 'tame' medieval death, as described by Ariès.[51] By the same token, a sudden and unexpected death was the worst death and could compromise the spiritual afterlife of the individual, even when they had lived a good life: Ariès cites a number of medieval sources that interpret sudden death as the mark of a curse.[52] Thus, although knowing the exact time and place of one's death might sound frightening to a modern sensibility, to a medieval mind it was a state to be hoped and prayed for.

Criminal death then had some important characteristics which gave it redemptive potential:

- It was scheduled and could thus be prepared for;
- The suffering of the earthly body could directly redeem some of the necessary pains of Purgatory;
- Analogy with the suffering body of Christ in passion and with tortured and mutilated saints' bodies made the interpretation of criminal death 'perilous'.[53] Regulated violence was 'sanctified' because suffering was part of God's plan.[54]

Pain: The Aim of Punishment or Its By-Product?

If death, even shameful and painful death by public execution at the hands of the State, could be reimagined as a holy path to redemption, was execution also expected to act as a deterrent? There is evidence from both the early and late medieval periods to suggest that it was, though it may be the case that the meaning taken by witnesses of an execution was never entirely within the control of the State, and that alternative, possibly subversive, parallel meanings could not be suppressed, given the pervasive symbolism of holy passion.

First, although it might have been the fate of one's soul after death that was the frightening prospect, rather than death per se, medieval executions were frequently painful and horrible deaths. Bodily pain was an element in many punitive sanctions in the medieval period, including whipping, or the removal of a hand, ear or another body part. However, inflicting pain was not such a central element of medieval punitive regimes as is sometimes imagined. As we saw earlier, and even in the case of strangulation hanging, branding, flogging, dismemberment or enduring the stocks or rough music, other aspects of those corporal and capital sentences such as humiliation or the bestowing of an enduring social stigma, were probably just as significant as pain in making the punishment fearful. Violence in punishment, therefore, was a necessary part of maintaining the social order, but its employment was always controlled, ordered and licensed, rather than being used for its own sake or in a way that might risk destabilising the social order.[55]

These other elements, though, were effective in evoking dread in most medieval minds. Public shaming and dramatic exclusion from the community were more important than pain, which was often incidental to the punishment. The main purpose of removing a hand, for example, was to render the criminal always visible and to mark their deviancy permanently and inescapably on their body. Such a procedure made full reintegration as a respectable member of the group all but impossible. Similarly, the memory of having been bound in the stocks or paraded through the streets endured long after the cuts and bruises had gone. Social exclusion was a very powerful sanction, especially in the early and High Middle Ages. In our modern age of quick and easy travel, voluntary emigration and reliable communication, leaving one community and joining another does not seem like a punishment. However, like a sentence of transportation in the eighteenth century, a sentence of banishment or exile in the medieval period was almost equivalent to death.

Reynolds notes that in the Anglo-Saxon period the clear distinction between 'members' and 'others' in society was maintained through geographical segregation as well as bodily practice.[56] Westerhof adds that whether by exile or excommunication, erasure from society was a dreaded fate, and cites John of Salisbury's observation that exclusion from society during life does not end at death.[57] 'Strangers' occupied lowly and disadvantaged positions in society, and could be excluded even from normal burial places.

Given the dread of being excluded from the community of the saved, it might seem surprising that medieval human justice did not try to impede the souls of notorious criminals from finding redemption. In fact, on the contrary, they seem to have been given every opportunity to save their souls: a scheduled time of death and provision of a priest to make confession: in short, the chance to die an exemplary death with prayer and penitence. Foxhall Forbes notes that some legal codes advocated giving enough time between sentence and execution so that the convict had the opportunity to express true penitence and ask for God's forgiveness, as well as to begin their penance.[58] This comes from another important and largely implicit cornerstone of medieval justice; that ultimately punishment is decided by God. Until the twelfth century, God's supremacy over human judgement was evident in the general practice of trial by ordeal. When an accusation was brought against a person, rather than attempting to enquire into the details of the evidence or the fairness of the accusation, the whole question was turned over to God. Ordeals might use cold or hot water, hot iron or armed combat to manifest the will of God. All were preceded by a period of prayer and spiritual cleansing. The ordeal by cold water involved submersion of the accused in a body of water, analogous to baptismal water which would embrace (i.e., pull under) the innocent and pure of soul, but float the impure. An accused person undergoing the hot water ordeal had to retrieve an object such as a stone from the bottom of a cauldron of boiling water. Like the ordeal by iron, which involved carrying a red-hot iron bar a distance of nine feet, divine judgement was manifested in how the wounds healed. If, after being bandaged for a few days, the scalded or burned flesh had recovered cleanly then the accused was innocent; a festering wound was an indicator of guilt. Ordeal by combat was, as it sounds, the will of God made manifest in a fight between the accuser or the defendant or their champions. The replacement of trial by ordeal with trial by jury was one of the conditions of the Magna Carta of 1215, and earlier in the thirteenth century, King John had tried to force a treason trial to be decided in gladiatorial combat between his own nominee and that of the Poitevin barons who, in an era-defining act of resistance, refused to recognise any other kind of trial than peer jury.[59]

The role of ordeal in British medieval punishment shows two things. First, the lack of distinction between sin and crime—God was to be the ultimate judge of both, and the role of the earthly judicial Establishment, like the role of the Church, was merely to control and operate structures in which the will of God could be exercised. Second, the ordeal shows how crime, like sin, was written into the material substance of the body. The body's buoyancy in water, its ability to heal from injury or prevail in combat was dependent on its spiritual state, which, in turn, was determined by the nature of unatoned sins or crimes carried by its soul. Not only was God's omniscience thus harnessed to resolve questions of guilt, but the ordeal enabled the process of punishment/penance to begin. The ordeal 'asked God to reveal guilt in the body so that the soul may be saved'.[60]

MEDIEVAL CRIMINAL LAW AND SANCTIONS ON THE BODY

Despite the stereotype of medieval punishment being brutal and bloody, as discussed above, many crimes in the Middle Ages were punished in other ways, particularly through the payment of fines or the forfeiture of estates. Even those found guilty of treason could often escape with their lives in the period before the fourteenth century, provided they were willing to make an apology, swear loyalty to the monarch, and forfeit all or a large part of their estate.[61] Banishment and exile were also common punishments for serious crime in the period, though they seem to have lost some of their sting by the late Middle Ages. Exclusion from the community appears to have been a particularly harsh fate in the early medieval period, and this is significant in understanding the symbolic importance of the Anglo-Saxon execution cemetery.

Powerful Punishments and Traitors' Bodies

So, was the power of the medieval criminal body harnessed? The answer is that it was—both as a material lesson in the power of the State and for its inherent potency. However, the first of these purposes was never unambiguously successful, as we shall see.

The power of the State was manifest most clearly in the criminal body of the traitor. More than any other crime (except perhaps suicide), treason was an affront to the natural order of God and man. The majority of those sentenced for treason were of aristocratic birth, these being the only people, as a rule, who had the status and resources to mobilise effectively against the monarch. Their crime was compounded by their failure to 'live up to the standards of aristocratic identity and community'.[62] Because aristocratic identity was realised through the practice and appearance of an ideal, highly gendered, aristocratic body, so too their 'dishonour was rendered visible within and upon the body'.[63]

The 1352 Statute of Treason formalised existing customary jurisprudence and practice. Particularly during the late medieval period the punishment of the traitor's body was a highly symbolic restitution of the social and divine order. Until the late Middle Ages treason was punished with 'a remarkable degree of clemency'.[64] No earl was executed for treason in England between the death of Waltheof in 1076 and that of John, Earl of Atholl in 1306; only direct attempts on the king's life were always punishable by death. In 1238 an *armiger literatus* (sergeant at law) was given the traitor's death of drawing, hanging, beheading, and quartering, and in 1242 William de Marisco was drawn, hanged, disembowelled and quartered, for example.[65] Both of these men had threatened the life of the king and thus the authority of God, since the king ruled by divine order. By the start of the fourteenth century the definition of treason had expanded to include offences such as making false coin and, in 1278, 293 Jews were executed in London for coin clipping in London.[66] However, the death of a single traitor in 1305 occasioned far more comment at the time and ever since. William Wallace was one of the leaders of a sustained Scottish revolt against Edward I in the late thirteenth century. After their eventual defeat, most other Scottish leaders agreed to the king's terms, and were granted a pardon in exchange for forfeiture of their estates. Wallace, however, refused to acknowledge the authority of the English king and was therefore punished very severely. Edward I appears to have directed particular enmity towards Wallace, perhaps because of his sustained defiance to the English king's claim to rule Scotland. In any case, he was not given a proper trial or the opportunity to defend himself after his capture in 1305. According to the chronicles of the time, William Wallace was drawn 'at the horse's tail' to the place of execution where he was hanged, but not to death. He was then taken down and beheaded. His entrails were removed and burned and his remains quartered and sent to Newcastle, Berwick, Stirling and Perth. His head was put on a spike and fixed to London Bridge.[67]

All the elements of Wallace's punishment had symbolic meaning. To be drawn to execution on a sled or hurdle was the mark of a traitor. The crowd that witnessed the procession of shame could augment this part of the punishment with jeers and missiles, performing the process of rejection and exclusion from the community of the faithful. Hanging alive, noted the author of the Dunstable annals in relation to the execution of Dafydd ap Gruffydd 22 years earlier, is the punishment for those who had killed men of high rank.[68] He was beheaded because of his outlaw status, and disembowelled because it was in his entrails that his acts of blasphemy were generated.[69] Dismemberment was the price of sedition and also allowed the deterrent effect of public display to work at several locations of treasonous activity. The northern towns to which Wallace's quartered body was sent were selected because of their significance in his rebellion. His head remained in London, the metaphorical head ('capital') of the country. In the years following the death of Wallace, a number of other Scottish rebels were also executed for treason. While these deaths generally followed the same pattern as Wallace's, there were some variations. Disembowelling could occur before beheading so that, in the worst cases, convicts would see their own entrails burning before they lost consciousness or died.

The symbolic elements of Wallace's trial execution were augmented by those who placed on his head a chaplet of laurel (or, in some accounts, of oak), in mockery of the crown he once claimed he would wear (though he did not claim the throne of Scotland for himself).[70] While intended to humiliate the body through parody of kingly regalia, it must surely have increased the resemblance of Wallace's ignoble end to that of Jesus Christ, wearing his crown of thorns in another parody of kingship. Because this representation of the body of the dying Christ was so extremely well-known and ubiquitous at the time, to crown Wallace with vegetation must surely have been an ideological own goal. Given that the traitor, Earl Waltheof, executed in 1076, was within a few years the subject of a cult of saintly veneration, the State might have realised that playing with the polyvalent symbolism of execution was a dangerous game. However, a few years later, Hugh le Despenser the Younger, executed for treason in 1326, was also made to wear a symbolic crown, this time of nettles. Musson suggests that the choice of plant might relate to heresy or be an indication that he had 'stung' people, but nettles, like Christ's crown of thorns, might have had no meaning beyond the ironic subversion of a hoped-for real crown into an ornament that would only add to his bodily suffering.[71] Despenser was also robed in a tabard with his family arms reversed, to signify the dishonour his treason had brought on his name. In other cases, a servant executed for a serious crime might be hanged wearing his master's livery.[72] In various ways, then, the bodies of criminals might be elaborated with clothing or ornamentation in order to clarify the symbolic meaning of their execution.

The geography of execution and its aftermath was also symbolically freighted. The recurrent use of traditional locations for the punishment of traitors and the display of their remains were meaningful in their own right and gained additional weight by repeated use. Traitors were usually tried and executed in London—a capital city for a capital offence—and specifically at Tower Hill, in the most secure and loyal heart of royal power. It is interesting that during the Peasants' Revolt in 1381, 'traitors' to Wat Tyler's cause were also beheaded at Tower Hill, both to mimic the judicial spectacle of power and to appropriate it.[73]

The overt symbolism of medieval bodily punishment was not limited only to executions for treason. Merback notes that a German sentencing formula specified that criminals executed by hanging should be suspended high up, using a new rope, and then left on the gallows for some time, 'so that it shall be given over to the birds in the air and taken away from earth so that furthermore neither persons nor property may be damaged by this man'.[74] The criminal corpse in this understanding is a source of moral pollution, and its emblematic nature demonstrates how judicial process occurred at the 'crossroads of law and belief'.[75]

Although post-mortem punishments did not become formalised until the Early Modern period in Britain, it is clear that already in the Middle Ages there were degrees of execution. A death penalty could be made 'worse' by the addition of extra elements of bodily suffering, but more commonly, the particular execution was given a more precise and nuanced meaning through the addition of connotative or moral elements that varied with the nature of the crime and the status of the criminal. These elements had a role to play in the restoration of society and the rebalancing of the social and spiritual disordering occasioned by crime.

For medieval people, the distinction between secular crime and religious sin was not blurred and often not meaningful. The process of penance and absolution for sin was continuous with the process of punishment for crime.

INTO MODERNITY

At the time that the medieval period segued into early modernity in the sixteenth century, the criminal corpse was already a significant symbolic locus which could be made to act as moral lesson, and a tool of State authority, or a source of medical and magical healing. However, the two post-mortem treatments that dominated the core period of the Murder Act—anatomical dissection and hanging in chains—were not part of the punitive repertoire. Although the punishment of treason had already developed the characteristics it would retain for the next few centuries, the aggravation of execution by the strategic and brutal use of pain was not yet widely practised, and pain in medieval punishment was sometimes incidental to the emphasis of a symbolic point. The criminal body—dying and dead—in the medieval period was an ambiguous thing at best. Because of the ubiquity of religious iconography featuring the suffering of Jesus Christ and the saints, the sanctifying and spiritual nature of physical punishment was inseparable from the secular judicial elements. Moreover, the division between demonstrative political uses of the criminal body, and the Purgatorial atonement for sin was blurred, if not meaningless at this time.

As the Tudor period began, England moved into early modernity. The religious upheavals of the fifteenth century would see the end of Purgatory for Protestants and a shift in the relationship between living and dead. The meanings of the body—and especially of the dead body—were altered in ways that seem in some ways unexpected, and this had an effect on the uses of the criminal corpse. As we shall see, as the medieval became the modern, punishment of the body became rather more brutal, and the suffering body was universalised less by the suffering Christ and more by the emergence of a new discourse of modern medical science.

NOTES

1. Tarantino, Q. (1994), *Pulp Fiction: A Quentin Tarantino Screenplay* (New York: Hyperion), quote at p. 108.
2. See, Foucault, M. (1977), *Discipline and Punish: The Birth of the Prison* (Harmondsworth: Penguin).
3. See, Dean, T. (2001), *Crime in Medieval Europe 1200–1550* (Harlow: Longman), pp. 119–120.
4. See, Bellamy, J.G. (1998), *The Criminal Trial in Later Medieval England: Felony Before the Courts from Edward I to the Sixteenth Century* (Stroud: Sutton), p. 57.

5. See, Wormald, P. (1988), 'A Handlist of Anglo-Saxon Lawsuits', *Anglo-Saxon England*, Vol. 17, 247–281, referenced in, Reynolds, A. (2009), *Anglo-Saxon Deviant Burial Customs* (Oxford: Oxford University Press), p. 10.
6. See, Ariès, P. (1974), *Western Attitudes Towards Death from the Middle Ages to the Present* (Baltimore: Johns Hopkins); Ariès, P. (1981), *The Hour of Our Death* (Harmondsworth: Peregrine).
7. For a critique of the term 'deviant burial', see Aspöck, E. (2008), 'What Actually Is a 'Deviant Burial'? Comparing German-Language and Anglophone Research on 'Deviant Burials'', in Murphy, E.M. (ed.), *Deviant Burial in the Archaeological Record* (Oxford: Oxbow Books), pp. 17–34.
8. See for example, Reynolds, A. (2009), *Anglo-Saxon Deviant Burial Customs* (Oxford: Oxford University Press); Murphy, E.M. ed. (2008), *Deviant Burial in the Archaeological Record*, Vol. 2 (Oxford: Oxbow Books); Cherryson, A.K. (2008), 'Normal, Deviant and Atypical: Burial Variation in Late Saxon Wessex, c.AD 700–1100', in Murphy, E.M. (ed.), *Deviant Burial in the Archaeological Record* (Oxford: Oxbow Books), pp. 115–130; Buckberry, J. (2008), 'Off with Their Heads: The Anglo-Saxon Execution Cemetery at Walkington Wold, East Yorkshire', in Murphy, E.M. (ed.), *Deviant Burial in the Archaeological Record* (Oxford: Oxbow Books), pp. 148–168. Not all deviant burials are criminal burials: there are many other reasons why an individual might be given unusual mortuary treatment, including circumstances of death, ethnic or other identity, religion, belief, being a victim of crime, a stillbirth or neonate, a casualty of war, a religious sacrifice or because their life or death made them more likely to become a revenant or to trouble the living. By the same token, not all criminals were necessarily distinguished by non-normative burial practices. In many circumstances, we have no way of knowing archaeologically whether a body is an executed criminal or not. A woman who had been judicially drowned and then interred in the normal local burial ground, for example, would be indistinguishable from the rest of the community in death.
9. See, Reynolds, A. (2009), *Anglo-Saxon Deviant Burial Customs* (Oxford: Oxford University Press), p. 5, pp. 61–95.
10. Meaney and Hawkes (1970), pp. 31–32.
11. See, Reynolds, A. (2009), *Anglo-Saxon Deviant Burial Customs* (Oxford: Oxford University Press), p. 234.
12. See, Foxhall Forbes, H. (2013), *Heaven and Earth in Anglo-Saxon England* (Farnham: Ashgate).
13. See, Harris, O.J.T. and Robb, J. (2013), 'The Body and God', in Robb, J. and Harris, O.J.T. (eds.), *The Body in History: Europe from the Palaeolithic to the Future* (Cambridge: Cambridge University Press), p. 136.

14. See, Mills, R. (2006), *Suspended Animation: Pain, Pleasure and Punishment in Medieval Culture* (London: Reaktion books).
15. See, Harris, O.J.T. and Robb, J. (2013), 'The Body and God', in Robb, J. and Harris, O.J.T. (eds.), *The Body in History: Europe from the Palaeolithic to the Future* (Cambridge: Cambridge University Press), pp. 129–163, p. 132.
16. Most recently, Tarlow, S. (2010), *Ritual, Belief and the Dead in Early Modern Britain and Ireland* (Cambridge: Cambridge University Press); Robb, J. and Harris, O.J.T. (2013), *The Body in History: Europe from the Palaeolithic to the Future* (Cambridge: Cambridge University Press).
17. See, Robb, J. and Harris, O.J.T. (2013), *The Body in History: Europe from the Palaeolithic to the Future* (Cambridge: Cambridge University Press), pp. 136–137.
18. For a discussion of somatophobia in the early modern period see for example, Berger, H. (2000), 'Second World Prosthetics: Supplying Deficiencies of Nature in Renaissance Italy', in Erickson, P. and Hulse, C. (eds.), *Early Modern Visual Culture: Representation, Race and Empire in Renaissance England* (Philadelphia: University of Pennsylvania Press), pp. 98–147.
19. See, Horrox, R. (1999), 'Purgatory, Prayer and Plague: 1150–1380', in Jupp, P. and Gittings, C. (eds.), *Death in England: An Illustrated History* (Manchester: Manchester University Press), pp. 90–118, quote at p. 93.
20. Ibid., p. 93.
21. Ibid., p. 100.
22. See, Park, K. (1995), 'The Life of the Corpse: Division and Dissection in Late Medieval Europe', *Journal of the History of Medicine and Allied Sciences*, Vol. 50, Issue 1, 111–133.
23. See, Bynum, C.W. and Freedman, P. eds. (2000), *Last Things: Death and the Apocalypse in the Middle Ages* (Philadelphia: University of Pennsylvania Press), p. 6.
24. See, De Wilde, P.M. (1999), 'Between Life and Death: The Journey into the Other World', in DuBruck, E. and Gusick, B. (eds.), *Death and Dying in the Middle Ages* (New York: Peter Lang), pp. 175–187, p. 175.
25. See for example, Bradford, C.A. (1933), *Heart Burial* (London: George Allen & Unwin); Weiss-Krejci, E. (2010), 'Heart Burial in Medieval and Early Post-Medieval Central Europe', in Rebay-Salisbury, K., Sørensen, M.L.S., and Hughes, J. (eds.), *Body Parts and Bodies Whole: Changing Relations and Meanings* (Oxford: Oxbow Books), pp. 119–134.
26. See, Geary, P.J. (1994), *Living with the Dead in the Middle Ages* (Ithaca: Cornell University Press); Walsham, A. ed. (2010), *Relics and Remains Volume 13* (Oxford: Oxford Journals).

27. See, Binski, P. (1996), *Medieval Death* (London: British Museum Press), p. 65.
28. See, Westerhof, D. (2008), *Death and the Noble Body in Medieval England* (London: Boydell and Brewer).
29. Ibid., p. 31.
30. See, O'Connor, M.C. (1942), *The Art of Dying Well: The Development of the Ars Moriendi* (New York: Columbia University Press).
31. Ibid.; Appleford, A. (2015), *Learning to Die in London, 1380–1540* (Philadelphia: University of Pennsylvania Press).
32. See, Davies, R.T. ed. (1964), *Medieval English Lyrics: A Critical Anthology* (London: Northwestern University Press), p. 74.
33. See, Joynes, A. (2001), *Medieval Ghost Stories: An Anthology of Miracles, Marvels and Prodigies* (Woodbridge: Boydell).
34. Ibid., pp. 97–98, 121.
35. Ibid., pp. 37–38.
36. Ibid., p. 35.
37. See, Gilchrist, R. (2008), 'Magic for the Dead? The Archaeology of Magic in Later Medieval Burials', *Medieval Archaeology*, Vol. 52, Issue 1, 119–159, quote at p. 120.
38. Ibid., p. 140.
39. See, Le Goff, J. (1984), *The Birth of Purgatory* (Chicago: University of Chicago Press).
40. See, Merback, M.B. (1999), *The Thief, the Cross, and the Wheel: Pain and the Spectacle of Punishment in Medieval and Renaissance Europe* (London: Reaktion Books), p. 150.
41. T. Olsen 2005. 'The Medieval Blood Sanction and the Divine Beneficence of Pain: 1100–1450'. Bepress Legal Series Paper 1334. (http://www. academia.edu/1339016/The_Medieval_Blood_Sanction_and_the_ Divine_Beneficence_of_Pain_1100_-_1450).
42. Jung, J.E. (2000), 'From Jericho to Jerusalem: The Violent Transformation of Archbishop Engelbert of Cologne', in Bynum, C.W. and Freedman, P. (eds.), *Last Things: Death and the Apocalypse in the Middle Ages* (Philadelphia: University of Pennsylvania Press), pp. 60–78, quote at p. 68.
43. Ibid., p. 71.
44. Ibid., p. 78.
45. See, Daniell, C. and Thompson, V. (1999), 'Pagans and Christians: 400–1150', in Jupp, P. and Gittings, C. (eds.), *Death in England* (Manchester: Manchester University Press), pp. 65–89, quote at p. 83.
46. See, Merback, M.B. (1999), *The Thief, the Cross and the Wheel: Pain and the Spectacle of Punishment in Medieval and Renaissance Europe* (London: Reaktion Books).
47. Ibid., p. 19.

48. See, Binski, P. (1996), *Medieval Death* (London: British Museum Press), quote at p. 47.
49. Passion plays are still regularly performed around the world, especially at Easter. The most recent fatality was not a Jesus, but a Brazilian actor playing Judas who was accidentally hanged when his safety equipment failed. For a discussion of 'passion plays' in the late medieval period see DuBruck, E. (1999), 'The Death of Christ on the Late Medieval Stage: A Theatre of Salvation', in DuBruck, E. and Gusick, B. (eds.), *Death and Dying in the Middle Ages* (New York: Peter Lang), pp. 355–370.
50. See, DuBruck, E. and Gusick, B. eds. (1999), *Death and Dying in the Middle Ages* (New York: Peter Lang), pp. 20–21.
51. See, Ariès, P. (1981), *The Hour of Our Death* (Harmondsworth: Peregrine).
52. Ibid., pp. 10–12.
53. See, Morgan, P. (1999), 'Of Worms and War: 1380–1558', in Jupp, P. and Gittings, C. (eds.), *Death in England* (Manchester: Manchester University Press), pp. 119–146, quote at p. 124.
54. See, Maddern, P. (1992), *Violence and Social Order: East Anglia 1422–1442* (Oxford: Clarendon Press), quote at p. 82.
55. Ibid., p. 115.
56. See, Reynolds, A. (2009), *Anglo-Saxon Deviant Burial Customs* (Oxford: Oxford University Press).
57. See, Westerhof, D. (2008), *Death and the Noble Body in Medieval England* (London: Boydell and Brewer), pp. 16–17.
58. See, Foxhall Forbes, H. (2013), *Heaven and Earth in Anglo-Saxon England* (Farnham: Ashgate), p. 185.
59. See, Fryde, N. (2001), *Why Magna Carta?: Angevin England Revisited* (Münster: LIT Verlag), p. 164.
60. Ibid., p. 200.
61. See, Bellamy, J.G. (1970), *The Law of Treason in England in the Later Middle Ages* (Cambridge: Cambridge University Press).
62. See, Westerhof, D. (2008), *Death and the Noble Body in Medieval England* (London: Boydell and Brewer), quote at p. 96.
63. Ibid.
64. See, Bellamy, J.G. (1970), *The Law of Treason in England in the Later Middle Ages* (Cambridge: Cambridge University Press), quote at p. 23.
65. Bellamy, *The Law of Treason*, p. 23.
66. Joseph Jacobs, 'England', *Jewish Encyclopedia*, 1901–1906.
67. The execution of William Wallace is related in, Westerhof, D. (2008), *Death and the Noble Body in Medieval England* (London: Boydell and Brewer), pp. 97–100; Bellamy, J.G. (1970), *The Law of Treason in England in the Later Middle Ages* (Cambridge: Cambridge University Press), pp. 33–39.

68. See, Bellamy, J.G. (1970), *The Law of Treason in England in the Later Middle Ages* (Cambridge: Cambridge University Press), p. 26.

69. The head of an outlaw—the 'wolf's head' that meant that an outlaw could be beheaded on sight with impunity—seems to have had particular symbolic resonance in the Middle Ages. See, Musson, A. (2001), *Medieval Law in Context: The Growth of Legal Consciousness from Magna Carta to the Peasants' Revolt* (Manchester: Manchester University Press), p. 20.

70. See, Stevenson, J. ed. (1841), *Documents Illustrative of Sir William Wallace: His Life and Times* (Printed for the Maitland Club), p. xxxiii.

71. See, Musson, A. (2001), *Medieval Law in Context: The Growth of Legal Consciousness from Magna Carta to the Peasants' Revolt* (Manchester: Manchester University Press), p. 232.

72. See, Bellamy, J.G. (1998), *The Criminal Trial in Later Medieval England: Felony Before the Courts from Edward I to the Sixteenth Century* (Stroud: Sutton), p. 154.

73. See, Musson, A. (2001), *Medieval Law in Context: The Growth of Legal Consciousness from Magna Carta to the Peasants' Revolt* (Manchester: Manchester University Press), pp. 246–247.

74. See, Merback, M.B. (1999), *The Thief, the Cross and the Wheel: Pain and the Spectacle of Punishment in Medieval and Renaissance Europe* (London: Reaktion Books), quote at p. 139.

75. Ibid.

CHAPTER 3

How Was the Power of the Criminal Corpse Harnessed in Early Modern England?

THE CONTEXT

The transition from medieval to early modern, though by no means clear cut, is a useful heuristic because of the deep intellectual and social changes that unfolded over the period of the Reformation and Renaissance. Dyer has noted that many of the processes traditionally identified with the Early Modern Period in fact have much earlier origins, and in this contention he fits into an established tradition of finding medieval origins of the modern world.[1] However, in the history of death in Britain (though perhaps not so much in the history of punishment), a sixteenth-century transition from medieval to modern, in which the Reformation plays a significant part, is a defensible division. The doctrinal and liturgical changes of the Protestant Reformation had profound effects on the way people prepared for death, and on the relationship between the living and the dead.

Changes in Criminal Justice During the Age of Spectacular Punishment

During the sixteenth century, the development of a modern, scientific approach to the body intersected with an increasingly formalised, wide-ranging and state-run approach to the law, to institutionalise post-mortem anatomical dissection as a treatment for the criminal corpse.

© The Author(s) 2018
S. Tarlow and E. Battell Lowman, *Harnessing the Power of the Criminal Corpse*, Palgrave Historical Studies in the Criminal Corpse and its Afterlife, https://doi.org/10.1007/978-3-319-77908-9_3

In 1540, Henry VIII founded by charter the Company of Barber
Surgeons, giving professional legitimacy and formal recognition to what
had not always been an entirely respectable trade.[2] The same charter
included provision for the bodies of four executed felons per annum to
be made available to the Company, for anatomical dissection. In 1565
the Royal College of Physicians started giving anatomical lectures, which
continued to be a college fixture until the building's destruction in the
Great Fire of 1666. Also in 1565, John Caius obtained an annual grant
of two bodies to be delivered to Gonville College, Cambridge, and in
1626 Charles I allowed the Reader in Anatomy at Oxford to claim the
bodies of anyone executed within 21 miles of the city.[3] Because it was
important that the bodies for dissection were as fresh as possible, surgeons liaised directly with the sheriff, whose job it was to see that sentences were carried out.

However, by the 1690s the Barber-Surgeons were finding it harder to
claim bodies from the sheriffs. Official supply of executed criminals could
not keep pace with demand. By the time of the Murder Act of 1752 only
a very few of the subjects of anatomical dissection came through judicial channels and very few ended up in the universities of Oxford and
Cambridge, or at the Royal College. The numerous private anatomy
schools that sprang up like mushrooms in the early eighteenth century
were unscrupulous in their efforts to obtain the necessary specimens for
educational dissection.

Punitive dissection was, argues medievalist Katharine Park, a north
European practice in origin.[4] In Italy, she argues, a different cultural tradition in the Late Medieval and Early Modern Periods meant that the
body was less closely associated with the person. A dead body had far
greater symbolic and emotional impact in northern Europe where folk
practices and beliefs suggest that the body, even after death, retained
some of the essence and life force of the person (an idea explored later
in this chapter). The display of the corpse in Italian justice was an assault
on the memory or honour of the executed, not an assault on their personhood, claims Park.[5] While Park's contrast is perhaps overdrawn—
it is certainly the case that the display of heads and quarters above city
gates, for example, was an attack on the memory and honour of north
Europeans, too—the idea of residual personhood does have explanatory
force in explaining some traditional medical beliefs and, allied to some
wobbly theology, might have been significant in shaping public attitudes
to dissection.

As Sawday observes, penal dissection per se is a creation of the eighteenth century.[6] In Tudor and Jacobean times, bodies, especially those of traitors, were pulled apart, but this kind of public dismemberment was about avenging an injury to the sovereign, and not primarily about either public recompense or scientific education. Although the bodies of some executed felons were made available (in vanishingly small numbers) for anatomical research, this post-mortem fate was not part of the sentence, nor was there any sense in which the likelihood of meeting the anatomist's scalpel was linked to the depravity of the crime. Hanging in chains, however, in contrast to dissection, does appear frequently as an explicit part of sentencing before the Murder Act (in fact, gibbetings were more frequent in the decades immediately preceding the legal formalisation of the punishment than in the years following it, as we will see in Chapter 6). Hanging in chains, moreover, was particularly linked to certain kinds of serious crime. These were often murders or aggravated serious property crimes, such as highway robbery, robbery of the mail or smuggling, especially when another person was killed or seriously hurt. The gibbet also related strongly to the location of the crime, and was a punishment that was given particular force by being staged at the place where the crime was committed. In this, hanging in chains was similar to, and indeed often followed directly from, hanging at the scene of crime, a practice which was widely practised up to the eighteenth century, especially in parts of southern England, but was becoming unusual by the early nineteenth century.[7]

The Uses of the Dead in Early Modernity

What can you make from a dead body? The answers are numerous, both literal and metaphorical, and the criminal corpse was used deliberately in all these kinds of creations. You can use the dead criminal body to build scientific knowledge. Through methodical exploration of the body's interior, the basic principles of scientific biology, medicine and surgery (as opposed to a medicine built on the authoritative statements of classical authors) were discovered. The Early Modern Period was key in the history of anatomy, and the methodical and objective study of actual bodies, rather than texts, was essential in this process.

You can use the actual material substance of the body to make medicine for the living: From skull powder to 'mummy' made from the desiccated flesh of ancient bodies—or the rapidly dried meat of more recent ones; from drinking fresh blood to curing tumours with the touch of a hanged man's hand, the vital force that persisted in a fit, young body after life had been suddenly extinguished could be captured and channelled into the failing bodies of those still alive.[8] With a dead body you can make magic. In early modernity the distinction between science, medicine and magic was not always clear or relevant. How should we categorise the widespread practice of carrying teeth extracted from the skulls of the dead, as a safeguard against toothache, for example? In other cases, such as the spell apparently used successfully by John Dee to raise a man from the dead and make him act as Dee's personal servant, we would have no difficulty in seeing the dead body as a place of magic.[9]

Less literally, one can use a dead body—and especially a dead criminal body—to make a point or a statement. The spectacle of bodily punishment and of post-execution sanctions upon the body, its particular memorability, made the executed criminal body in early modernity an exceptionally potent symbolic resource in the construction of political power. A dead body can be used to make stories, or even to make history. By elaborating, venerating, desecrating or reshaping a body, it can be made into a saint or a villain according to the tales told about it. Whole or in pieces, the dead body provides the raw material that the storyteller, historian, politician or poet reconfigures to make a particular kind of truth.

All these uses for a dead criminal body had occurred in medieval England, but during early modernity they were more extensively discussed and formalised. The role of law in determining the treatment of an executed body was stronger and more consistent; the practice of scientific anatomy more methodically set out; and the curative powers of corpse medicine classified and codified in surgeons' and apothecaries' compendia and *vade mecums* (handbooks).

This chapter surveys the uses of the criminal corpse from the mid-sixteenth to the mid-eighteenth century. While the body during this period was not as extensively controlled by the State as it would be during the period of the Murder Act, it was nevertheless at this time that most of the legally determined treatments of the criminal corpse—hanging in chains, or dissection as punishment, for example—entered into customary use. Our approach to the criminal body in this chapter, and indeed in the book as a whole, is to view it as a powerful resource which could be deployed strategically to rhetorical or practical effect.

Changing Meanings of the Dead Body

Building a New Medicine

The sixteenth, seventeenth, and eighteenth centuries were periods of rapid progress in the exploration of the body and developing scientific understanding of anatomical systems and processes. The centuries between the European Renaissance and the age of the Murder Act were a period of transformation in how the human body was viewed, from a collection of humours to an integrated machine.[10] Over the same period, the dominant way of knowing and understanding the body also shifted: this time from a knowledge gained from books to one that accorded greater significance to first-hand observation and experiment. Knowledge of the world, gained from the new scientific principles, was also over this period transformed from being a way to know God, to being a way to change the world. To assert human mastery over nature one had to know it deeply, and with that knowledge came the ability to shape and control it. That power included the human ability to transform materials and substances, to change the shape and productivity of the land by mapping, enclosing, taming and improving it, and of other animals. It also included the ability to intervene in the workings and faults of the human body. That the body, once the soul had gone, was a part of Nature was increasingly taken for granted in the period. Harris, Robb and Tarlow describe the process by which new philosophy and theology converged on a new kind of mind/body split that 'provided a whole new way of looking at the world, a new kind of gaze'.[11] This new way of looking—investigative, empirical, neutral—was science, and it became the dominant approach to the human body.

The human body, like the universe of which it was a part, followed the rules and obeyed the mechanisms created by God. A new consensus developed in the early modern period about what knowledge was. The world could be better known through direct observation than through the writings of authorities, a position that mirrored the Protestant belief that the human relationship with God should be direct and mediated only by prayer and the Bible, not through the intermediaries of priests and teachers. Where possible, practical knowledge and experiment should form the basis of understanding, as opposed to medieval knowledge which paralleled practical experience of the world with an authoritative written discourse sanctioned by ancient authority and within which contradiction was not a problem. C.S. Lewis pointed out in his review of medieval bestiaries that the authority of literature was not only important in describing exotic animals such as giraffes and elephants, but also provided 'facts' about familiar domestic animals that must surely have been at odds with people's daily experience, such as the assertion that horses shed tears at the death of their master, or that the adder protected itself from snake charmers by curling up with one ear against the ground and the other stopped up with the tip of its tail.[12] Similarly the idea was advanced that beavers could cast off their genitals to distract predators while they escaped, hence its Latin name, *castor*.[13]

The new medicine involved a paradigm shift from respecting the word of authorities to learning directly from dissected cadavers. This transformation is generally located in sixteenth-century Europe and is attributed to Vesalius.[14] It was assisted by a general opening up of attitudes towards intellectual discovery and the eventual acceptance by Pope Clement VII of the practice of human dissection for anatomical purposes.[15]

In the case of the human body, the dominant medieval medical approach had seen disease as the result of an imbalance of humours, the four essential life fluids, following the teachings of Galen.[16] This humoral theory also made character inseparable from health. The empirical gaze of the new medical science replaced the sack of competing fluids with a balanced mechanism that needed to be observed. Overlapping both the humoral model and the machine, there was a period in the seventeenth century where one of the widespread metaphors of body was the microcosm—the body as geography whose uncharted waters and mysterious territories needed to be explored and mapped.[17] The consequent privileging of a rational, masculine, knowing subject, valorised as pioneer, explorer or hero—is at the heart of recent attacks on Enlightenment thought.[18]

Being a doctor or surgeon went from a trade to a profession and by the eighteenth century was a respectable life for a gentleman. Medical knowledge was collective and incremental, accumulating in books and periodicals, communicated through formal, college-based curricula, lectures and demonstrations.

The detailed mapping of the human body was fundamental to the project of understanding its internal relationships and mechanisms. A cartographic metaphor of exploration and geography was widely used to describe this undertaking.[19] As a metaphor of anatomy, the voyage of discovery sat equally well with a microcosmic view of the body, still important during the sixteenth and seventeenth centuries. It was also an attractive analogy for the new science because it conveyed the importance of actually visiting—seeing for oneself—the workings of the body. Etymologically, of course, the autopsy was a 'seeing for oneself' of the body's interior.

Revealing the complexity of the body's hidden and secret places was not, until the eighteenth century, fully divorced from a theological or philosophical consideration of the body. The introduction to Anthony Nixon's 1612 book of anatomy (significantly entitled *The Dignity of Man, Both in the Perfections of His Soule and Bodie*) answers the question, 'What commoditie cometh by anatomy of the body?' with the point that 'It puts us in minde of our mortality, and teacheth us that if the providence of God be so wonderful in the composition of the vilest and the earthly partes, It must needs follow that it is farre more great, and admirable in the creation of the Noble parts, especially of the Soule'.[20] Until the beginning of the eighteenth century, the introductions to most anatomical textbooks presented the detailed study of anatomy as a way of better knowing the self ('nosce te ipsum' being a recurrent motif in illustrations and frontispieces and thus of knowing God - Fig. 3.1).[21]

Of course, in order to admire the infinite subtlety and beauty of the mind of the Creator, direct, physical experience of the body's interior was essential. A supply of passive objects for the probing eye and scalpel of the anatomist had to be secured. Although only a comparatively small number of people in early modern England had direct and personal experience of anatomical dissection, the idea of anatomy and dissection was a potent cultural metaphor, explored further in Chapter 8.

The Dissected Body as Cultural Symbol

The anatomical dissection of the criminal body provided the structuring metaphor for numerous cultural phenomena in the early modern period. The separation, enumeration and description of body parts is evident in, as Hillman and Mazzio note, 'pictorial isolation, poetic emblazoning, mythic spargamos, satirical biting, scientific categorising, or medical anatomizing'.[22] Numerous early modern texts directed towards the exploration and rational examination of a topic called themselves 'anatomies'.[23] The metaphorical use of 'anatomy' for any analytical examination continued into the eighteenth century.

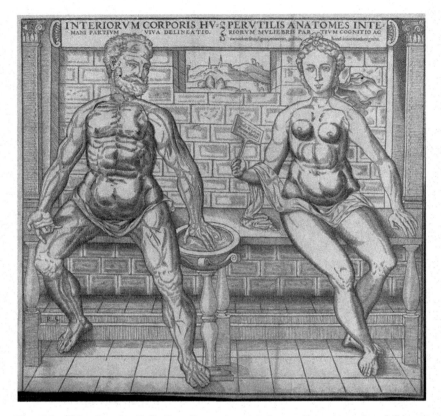

Fig. 3.1 Woodcut by R.S. *Interiorvm corporis hvmani partivm viva delineatio* (The Thomas Fisher Rare Book Library, University of Toronto)

Medicine and Folklore

In today's world, there is a wide choice of possible treatments for arthritis. We have little difficulty in categorising most of these as belonging to either orthodox, scientific medicine or alternative, folkloric practices. A doctor who has trained in medicine in most of the world might recommend a course of steroid injections or refer you for surgery, but is unlikely to suggest you carry a potato in your pocket. In the seventeenth century the separation between folk and orthodox medicine was less clearly established, and in many ways the period was one during which modern medicine tried to cut free from superstition and unsubstantiated beliefs based on now discredited models of health and disease. The survival of numerous folkloric remedies into the nineteenth century and beyond, when many of them were collected by county folklorists, seems to have been particularly evident among poor and rural people, who were presumably unable to afford trained, professional medical care.

The Social Consequences of Deviancy

The body of the executed criminal was capable of bearing a significant symbolic weight. Take, for example, the case of Guy Fawkes, an executed criminal whose proxy body still anchors a calendar festival that rehearses the social consequences of deviancy. As every British schoolchild knows, Guy Fawkes was one of the instigators of a Catholic plot to blow up parliament during its official opening, when James I would have been present, on 5 November 1605. The plot was discovered and Fawkes was found with 20 barrels of gunpowder in a cellar beneath the Houses of Parliament. Apparently spontaneous celebrations for the preservation of the king began that very year, with bonfires being built around London. An Act of parliament then ordered that the anniversary of the foiled plot should be celebrated annually as a day of thanksgiving for the king's life. In the next few years the practice of celebrating the anniversary of the occasion with bonfires spread around the country. Effigies of Guy Fawkes, and sometimes also the Pope, were customarily placed on the bonfire. Today, 'bonfire night' or 'Guy Fawkes day' is known and celebrated in most of Great Britain and in some colonial/post-colonial countries too. Burning the guy is the central and indispensable part of bonfire night tradition. Insider/outsider status is clearly enacted upon the body (or its proxy) in a way that not only expresses but also constructs social expectations.

The easy slippage between body and effigy was not unusual in early modern England, as we will see in the case of Oliver Cromwell, for example.[24] The body of Guy Fawkes, or rather a resurrected and recreated simulacrum or pastiche of his body, became the object of a ritual designed to reflect and create standards of political, religious and social conformity. However, in the twenty-first century, the executed traitor Guy Fawkes was again resurrected and transformed to do duty as a force of resistance and critique to government.[25]

The Reformation and the End of Purgatory

There is little consensus among historians of England about the nature and extent of change consequent upon the Protestant Reformation of the mid-sixteenth century. The redistribution of former Church property undoubtedly reshaped the political and economic as well as the geographical landscape. Max Weber's argument that a distinctively Protestant way of thinking permitted or even promoted the individualistic capitalism of modernity has been widely influential.[26] Protestantism does appear to be associated with the key social, aesthetic and philosophical viewpoints of the seventeenth and eighteenth centuries, although whether religious doctrine followed, informed or developed alongside broader attitudinal change is debatable. One area, however, in which religious reformation undoubtedly did lead to a thoroughgoing change in practice, discourse and feeling is in the relationship between the living and the dead. Although Ariès claims that differences between Catholics and Protestants are insignificant 'on the level of collective psychology', in fact the transformation from late medieval Catholicism to early modern Protestantism profoundly altered the parameters of our relationship with the dead.[27] It is possible that this change in turn affected the Catholicism of the Counter-Reformation on the continent. But to deny the impact of a new post-mortem geography is to ignore changes in secular as well as devotional practice. In the previous chapter, we considered the significance of Purgatory as an essential conceptual space, in which the living were able to interact with the dead. While the dead waited out their time in Purgatory, expiating their venial sins and preparing for salvation, their friends, relatives, beneficiaries and descendants could help them along with prayers, masses and gifts to the Church. For the living, who were thinking about their own mortality and the probable fate of their souls (and in the late Middle Ages, that meant pretty much everyone), the capacity of this world to have an impact upon the next meant that they could improve their own chances of Heavenly resurrection by endowing Church establishments in exchange for promises of prayers for their souls after death. There was also money to be made by the Catholic Church for selling indulgences, bulls and other 'Get out of Jail Free' tokens.

By getting rid of Purgatory the Protestant reformers utterly transformed the easy reciprocity between living and dead that the medieval economy of prayer and intervention had allowed. At death, a Protestant stood alone before God. Their salvation, the chances of which ranged across Protestant sects from fairly good to vanishingly slim, had already been determined, and not a million masses, nor all the indulgences in Europe could make a bit of difference. For the bereaved, this change left them powerless in any spiritual capacity. All that was left to them was to perpetuate the earthly fame of the dead, which they did through a new fluorescence of commemorative monuments, paintings and the emerging genre of memorial and mourning poetry.[28]

At the same time, the folklore of ghosts, monsters, revenants and fairies, always a little incoherent and contradictory, lost an important part of its geography: a place for 'all the disjecta of peripheral human experience'.[29] Since magical beings could no longer come from the moral Switzerland of Purgatory, it appeared to many that they must therefore come from Hell, and be unambiguously the Devil's cohorts. Periods of intense anxiety about the work of the devil, and associated waves of witch persecution punctuated the two centuries following the Reformation.

Protestants' rejection of the capacity of the living to influence the fate of the soul ironically led to an even greater concentration on the dead body. The bereaved began to channel the time, energy and resources they would formerly have invested in prayers for the soul into elaborate obsequies, enduring memorials and attention to the dead body itself. This was ironic because Protestantism shared with Catholicism a generally disdainful and suspicious attitude towards the body—living or dead. The body, including its functions, urges, and mutability was a temptation to sin and a source of wickedness. The living body must therefore be denied, ignored and opposed as far as possible. William Sherlock, a prominent seventeenth-century Protestant theologian suggests that, since the body was ultimately fated to decay, and bodily drives were temptations to sin, the living should endeavour:

> to live without our Bodies now, as much as possibly we can ... to have but very little commerce with flesh and sense; to wean our selves from all bodily pleasures, to stifle its appetites and inclinations, and to bring them under perfect command and government[30]

The body after death was an object lesson in vanity, futility and the inherent failure of the flesh. Another Protestant theologian, Zacharie Boyd, was in little doubt as to the ultimate value of the body:

Is it not your greatest desire to flitte from this bodie which is but a *Booth*, a *shoppe*, or *Tabernacle* of clay? Is not your Soule wearied to sojourne into such a reekie lodge?[31]

The *memento mori* tradition of the late Middle Ages used the image of the dead body (as rotting cadaver or dry bones) to emphasise the transience of human life and the inevitable fate of the flesh (Fig. 3.2). In the early modern period the dead body continued to function as a moral lesson to the living, reminding them to prepare for their own death (all the more urgent now that the preparation of one's own soul during life was the only way to improve one's prospects of salvation afterwards), but it was also a material demonstration of futility and of the unworthiness of the body. The dead corpse was body without soul. Its ugliness and stink showed the baseness of earthly life and proved that all beauty in the body had come from the soul.

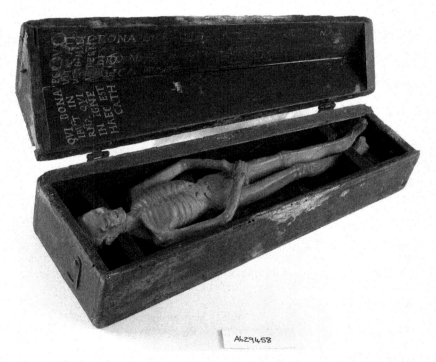

Fig. 3.2 Memento mori figure, a talisman to remind the holder of the transience of life and the vanity of earthly attachments (Wellcome Collection)

In this moral and religious climate, the body of the executed criminal was already freighted with meaning. Sin—in the form of acting on the criminal lusts of the unconstrained body—had resulted in death, and what an ugly, abject, dishonourable thing that traitorous body turned out to be when the spark of divinity was removed.

For condemned criminals, as for all people, the moment of death took on even greater significance as the state of the soul was sealed at this point. The last few minutes or even seconds of life became more decisive of a person's fate. Even a very late repentance, if sincere, could save the worst of criminals. This added drama to an already highly symbolic moment: would the sinner be damned or could they yet save themselves?

Ideal deaths were described in *ars moriendi* (art of dying) books, a literary tradition that began in the fifteenth century and continued to flourish through the early modern period.[32] The changing religious content of an ideal death, and the gradually shifting priorities of 'Moriens', the central character of the dying man, are a useful insight into how the Reformation and accompanying social changes impacted on what people aspired to achieve at the moment of leaving life. In a post-Reformation context, as in earlier periods, secular concerns such as writing a will and arranging for the payment of any outstanding debts, are dealt with quite quickly, and the real focus is on preparing the soul, through prayers of true contrition and repentance. Protestant *ars moriendi*, however, are distinguished from their predecessors by the absence of formalised ritual wording, and an additional emphasis on the hope of salvation and forgiveness, and the reduced role of friends and family, whose prayers no longer 'counted' for the dying, though they could still remind the person at the centre of the drama of their spiritual hopes and dangers.

Houlbrooke examines the profuse literature dealing with the good death that was published during the 150 years following the Reformation.[33] He points to some of the most popular *ars moriendi* books of the seventeenth to early eighteenth centuries, including John Hayward's *The Horrors and Terrors of the Hour of Death*, which had 21 editions between 1690 and 1730, and *Hell's Everlasting Flames Avoided*, published in 35 editions over a similar period. William Sherlock's *A Practical Discourse Concerning Death* went through at least 46 editions from 1689. In some ways, argues Houlbrooke, the moment of death was less rather than more important.[34] To a well-prepared soul in the Puritan tradition, the spiritual work necessary for salvation was accomplished during a virtuous and devout life; no particular death-bed performance was required. However, in the case of a bad life—which all condemned criminals had necessarily followed—a final moment of repentance could still be a redemptive act. Houlbrooke also notes, however, the power of a condemned criminal to subvert the expected rituals 'by a show of debonair indifference, or their drunken stupor at the gallows'.[35] Foreign observers, including Henri Misson, remarked on the fine appearance of those going to be hanged.[36] It was customary in early modernity for them to be well-dressed and newly shaved for their final journey. Gemelli, an Italian visitor, noted in 1701 that the condemned approached the scaffold 'as if going to a wedding'.[37] The comparison is apt. A popular belief was that a man could be saved from the noose if a woman agreed to marry him.[38]

In a survey of the popular murder literature of the early seventeenth century, Peter Lake identifies the tradition in pamphlets of crime and punishment that demanded a public confession and an expression of contrition before death.[39] This was not only a religious imperative, but also a theatrical restoration of moral and social order. Like Houlbrooke, Lake finds that the condemned criminal could, and sometimes did, subvert the prescribed behaviour. When the prisoner went to the scaffold drunk or ostentatiously unrepentant, the execution could become a carnival, exploiting the popular conventions of inversion and misrule. Like other carnivals of misrule, 'the executions... and the popular festivities which accompanied them, were structured by precisely the same principles of inversion followed by the reaffirmation of social unity'.[40] But where the usurping carnival king would be deposed in order to restore order, the criminal Lord of Misrule was actually rather than symbolically slain.

Popular murder narratives closed down ambiguity and imposed a particular interpretation, sometimes using the familiar arc through inversion of the normal order to a restoration and reaffirmation to channel the narrative to its moral.[41] The theatre of the scaffold, the gibbet and the anatomy room attempted the same kind of scriptwriting, using the actual body of the transgressor as both actor and prop.

As we shall see in the next section, attempts to tie down the criminal corpse to an unambiguous true story were not entirely successful. Counternarratives positioning the criminal as hero or martyr, for example, could be appropriated and subvert the orthodox account. Laqueur, writing about execution mostly in the eighteenth century, argues that the carnivalesque elements of execution, both intentional (the bravado of the central subject) and unplanned (ropes breaking, nooses failing, squabbles breaking out on the scaffold) meant that the State was never sure of being able to impose its lesson on its own terms.[42] We shall return to the disruptive potential of the polyvalent theatre of execution in later chapters.

Beattie says that in the seventeenth century there was a move away from capital punishment for serious crimes as the option of transportation became available.[43] This is certainly compelling if capital punishment is considered mainly as a solution to the problem of what to do with convicted felons. But a hanging—or even more, a beheading or burning—was not only, and maybe not even primarily, an efficient way of processing a deviant body out of society. The many ceremonial and public aspects of early modern execution demonstrate that a hanging accomplished symbolic ends that a simple removal from normal life and decent company, as with transportation or incarceration, could not. It was important that justice should be seen to be done: retribution achieved, social revenge enacted, would-be enemies of society deterred. While in this period, transportation could often be a lethal sentence, a death abroad or at sea did not have the theatrical and demonstrative potential of a staged execution.

Good and Bad Deaths

Despite the reformed Church's emphasis on the importance of soul and the insignificance of body, the 'decent' treatment of the corpse remained a priority for most. Perpetuation of a secular 'afterlife' replaced the spiritual one for survivors. The appropriate and dignified disposal of the corpse was important to everyone, and was afforded to all but the worst offenders. In 1739 Richard Tobin, condemned to hang for theft, wrote to his former master, 'Take some pity on me... for my friends is very poor and my mother is very sick, and I'm to die next Wednesday morning, so I hope you will be so good as to give my friends a small trifle of money to pay for a coffin and a shroud, for to take my body away from the tree that I am to die on'.[44] It is notable that by the early eighteenth century, a respectable burial, even for the very poorest in society, involved a shroud and coffin. Coffin use, a limited and elite practice for most of the Middle Ages, became by the eighteenth century part of the minimum requirement for decency. Laqueur notes that as the expectations of a decent funeral expanded, the lack of a full and proper one became more shameful.[45] By the nineteenth century, a burial that smacked of poverty was as shameful, he claims, as one that signalled criminality.

Prisoners who died in gaol were normally returned to their family and their home parish for burial. Most executed offenders were buried, although this was commonly carried out in the graveyard of the parish where the gaol or the scaffold was located.[46]

In early modernity, then, post-execution defilement gained its potency from its comparative infrequency and because it was a denial of what was most important. As Gittings comments, 'this desire to punish the dead corpse… makes sense only against a background of a society in which the decent interment of the dead was a matter of the utmost concern'.[47]

Honourable or dishonourable treatment of a dead body, even years after the death, not only reflected, but actively constructed a person as a respectable individual or a criminal. Two contrasting cases from early 1661 illustrate this. Following the restoration of the monarchy, the body of Oliver Cromwell, regicide, self-proclaimed Lord Protector and crypto-king, was disinterred from its resting place in Westminster Abbey, hanged at Tyburn and then beheaded. As was the case with others convicted of treason, Oliver Cromwell's head was placed on a spike and displayed to the public. This (delayed) post-mortem punishment stands in contrast to the ostentatious and hugely expensive state funeral he had been granted at the time of his death three years earlier.[48] At almost exactly the same time as Cromwell's body, along with those of fellow regicides Bradshaw and Ireton, was being dragged to Tyburn, the Marquis of Montrose underwent a similar reversal of fortune, though in his case the post-mortem transformation was from traitor to national hero. The Marquis had been executed in Edinburgh in 1650 when he had been fighting for Charles II against Cromwellian forces and dismembered. Eleven years later his remains were reassembled and brought together in a series of ceremonial processions involving velvet canopies and elaborate caskets for a lavish funeral, costing the enormous sum of £802 sterling, conducted at the king's expense.[49] In both these cases, the body itself was manipulated, even years after death, to make a particular story: in Cromwell's case it was to transform the story of a great ruler into one of a traitor, and an aberration in the history of England; in Montrose's, a criminal and traitor was transformed into a hero and martyr.

Uses of the Criminal Body

The Power of the State

One of the first major developments of early modern political history to be introduced to new students of the period is the formation of nation states. Some countries, including England, had already mostly coalesced during the Middle Ages (and others, such as Italy and Germany, took rather longer), and during this period larger, more unified and centralised political states took the forms that they held through modern history. As part of this process, the government of the state—whether monarchy or, increasingly, a more parliamentary institution—was at pains, in overt and subtle ways, to assert its authority. The consolidation of state control happened against a background of social reorganisation in the wake of Reformation, early English colonialism and the advent of sustained European transatlantic contact, and as an individualistic capitalism began to replace a totally local and cryptofeudal set of personal loyalties.

A key site in this renegotiation was the human body. The control and discipline of the living body was fundamental to running a modern State. An efficient army and a lively economy depended on work discipline, bodily knowledge, and rigid conformity with codes of physical behaviour, as has been extensively discussed by Michel Foucault and Norbert Elias, and before them by Max Weber.[50] The disciplined body was essential also in the controlled exchange of labour, materials, and goods which made the protocapitalism of early modernity.

The ideal body for the nascent nation state, then, was a controlled, disciplined, productive one. A high degree of conformity to shared standards of behaviour was encouraged by the state and facilitated by adherence to a national religion. Failure to live up to the normative standards of behaviour appropriate to a person's age, status and gender resulted in sanctions. For minor social transgression, the local 'moral economy' drew on a repertoire of punishments that mostly worked through shaming the malefactor and almost invariably centred on the humiliation of the body. These included medieval hangovers like branding, punitive amputation, and public display, such as a period in the stocks, public whipping or one of the various local manifestations of charivari—riding the strang, rough music, skimmington, *ceffyl pren*—all of which involved some kind of humiliating procession through town, usually on a wooden horse, in a state of undress or foolish attire, while enduring the catcalls and assaults of the community. Sanctions like these helped to police sexual behaviour, gender roles, and moral or religious obedience. For more grave crimes including serious violence and murder, major property crime, and especially anything that might be considered treason, the state intervened to impose a punishment decreed by an increasingly centralised and formalised judiciary.

As in the medieval period, nearly all punishments were public and depended on the twin forces of pain and shame to act as deterrent, retribution, or memorable display. The early modern period was the golden age of spectacular punishment. The body of the convict had a starring role in this pageant of retribution. On it—alive and dead, whole and in pieces—were piled humiliations and horrors for the consumption of the crowd. These were brutal demonstrations of state power and the price for opposing it. Linebaugh uses the word 'thanatocracy' (in much of his published work) to mean a government that rules by deploying the death penalty.[51] He interprets public hanging as a tool in the shaping and control of the working classes during the rise of capitalism.[52] But some criminals chose to go defiantly to their ends.

Subverting the Theatre of Pain

Even at its most authoritarian and repressive, the state always ran the risk that key performers in the spectacle of pain would go off-script and undermine the whole show. Because the criminal about to be executed was, first, likely by definition to be the kind of person willing to ignore or rework social norms and, second, given a person with little or nothing to lose at this point, the performance could go dangerously awry. The crowd had to be sensitive to the agony of the criminal, but if they identified with them too closely, or found the malefactor too charming or sympathetic, then they felt antagonism towards the state rather than awe. If the prisoner was amusing or drunk, or things went wrong with the actual execution, the solemnity of the occasion could be undermined by comedy. The ambiguity of the criminal facing death that was outlined in Chapter 2 persisted into this period. The line between felon and martyr; traitor and hero; murderer and saint could easily collapse if the protagonist refused to play the proper part. Ideally, condemned convicts met their fate with penitence and regret, acknowledging the justice of the sentence, as they acknowledged the ultimate power of the state as the guardian of orderly society. Such a role was salutary to the crowd, and the moral order of the nation was restored in the ceremonial payment of a blood price. In order to achieve the ideal death, one in which 'blood might be recompensed with blood and the land may be cleansed from the guilt thereof', the criminal should make a public confession and an expression of remorse.[53] Ministers were brought in to coach and prompt the principal actor, and executions were occasionally delayed if it seemed likely that a little more time would yield the desired result.[54]

The hope of an eleventh-hour pardon was reason enough for many criminals to work for a delay in execution. Laqueur notes that the frequent expectation of a last-minute reprieve produced a 'comedy of pardons' in the seventeenth century, where uncertainty about the outcome of a planned execution detracted from the dignity of the occasion.[55] For Laqueur the numerous contingencies of the execution event prohibit any reductive interpretation of the complex scene as 'coherent state theatre' in a Foucauldian sense.[56]

Laqueur's reservations are not uniquely retrospective. By the eighteenth century, literate commentators were expressing similar sentiments. For Samuel Richardson all the faces in the Tyburn crowd showed 'a kind of mirth', the thrill of pleasure at a public theatrical entertainment rather than fear and awe to see the might of the state in action.[57] Fellow novelist Henry Fielding set out a detailed account of his objection to public punishment.[58] For Fielding it was not the comedic element that tended to undermine the moral lesson, but the pathetic ones. The drama of the execution failed in its intent because the audience felt sympathy with the criminal as they faced the pitiless machine of justice. The perpetrator of violence became its victim. In other scenarios, repugnance for the grislier elements of physical or capital punishment undermined rather than reinforced the legitimacy of the state.

The punishment for treason remained the same as in the medieval period, and its elements equally symbolic. John Owen's sentence for treason in 1615 included first, that he be drawn to the place of execution 'as he is not fit to walk upon the earth: 2. His privy members cut off ... which shows that his issue is disinherited with the corruption of blood... 3. His bowels burned because in them he hatched the treason: 4. Beheaded: 5. Dismembered'.[59] The extremely somatic nature of both crime and punishment is very explicit here. But how far was a symbolic glossing of the components of punishment necessary or available for the attending crowd to interpret the event properly?

Even after death, the body of a traitor could be co-opted into new and subversive narratives. The family of William Stayley, a Catholic executed in 1678 for high treason, asked for the return of his body for burial. Because of his penitent behaviour before death, Charles II granted their request. However, instead of conducting a funeral of the decent but humble type consistent with Stayley's role as a repentant criminal, his family chose to bury him in the lavish and ostentatious fashion of a hero and martyr. The king was not prepared to lose control of the narrative. He ordered the body to be disinterred, and Stayley's quarters to be set up above the gates of London as originally planned.[60]

In the cases of other Catholic martyrs (and the events of the sixteenth and seventeenth centuries meant that there were dozens of them in the early modern period), body parts removed as part of a punishment that worked through humiliation and the denial of a decent burial were reappropriated and redefined as holy relics. The hand of St Margaret Clitheroe, executed in 1586 is still preserved and venerated at the convent of the Blessed Virgin in York; and that of St Edmund Arrowsmith, executed in 1628 is at St Oswald's church, Ashton in Makerfield. The body of another Catholic saint, St Oliver Plunkett, who was executed for treason in 1681 was rescued from the pyre after death, divided and preserved as holy relics in several locations around Europe.[61] Most famously, his head, once held up as that of a traitor, is now the most important relic at St Peter's Cathedral, Drogheda. His head has participated not only in religious narratives of faith and martyrdom, but also nationalist ones of oppression and resistance.

The creation of a martyr—either an actual religious martyr or a social and cultural one—involved not only material relics, but also the invention and propagation of stories. The cultural afterlives of criminal bodies will be discussed further in Chapter 8.

Early Modern Criminal Bodies

Archaeologists in general derive a great deal of their evidence about the past from the remains of the dead. Bones and graves, after all, leave material traces that frequently survive and that can be interrogated to yield information about past practices, including diet, social organisation, cultural affiliation, social identities of gender, age and status, disease, consumption and economics, among other things.

Accordingly, a commonplace has arisen that when we study the dead we are actually studying the living. The dead, goes the archaeological wisdom, do not bury themselves. There has thus been an assumption in much archaeological work that the bodies of the dead are passive and manipulable symbols through which the living pursue their own strategies of representation that promote the social or economic interests of their own kind. A study of the criminal corpse in early modernity exposes problems with that perspective. First, such a view takes no account of the emotional impact of death, especially a stigmatised, traumatic and early death such as execution. For the executed person, for their kin and for the wider community of witnesses, execution was a terrible fate, and could inspire fear, anger, sympathy, grief, disgust, awe, contrition, spiritual reflection, revolution or any combination of feelings. Second, the dead criminal body was an unreliable symbol that was not easily co-opted into the meaningful narrative composed by any group or individual. Instead it was polyvalent and ambiguous. The living may bury the dead, but the dead resist the stories we impose upon them.

The Criminal Body in Different Belief Discourses

At the outset of the project on which this book is based some colleagues were concerned that executed criminal bodies were too few in number and too marginal to the experience of most people in the past to make a sufficiently fruitful focus for a large interdisciplinary endeavour. In fact, not only have our findings exceeded our expectations, the interdisciplinarity of our work has allowed some new things to emerge that would never have arisen if we had followed separate traditional disciplinary paths. One of these is the issue of how different belief discourses relate to one another. The criminal body had a prominent presence in popular culture as well as science, civic life and medico-legal activity. It is historically significant as the site of overlapping and sometimes contradictory understandings between scientific anatomy, criminal justice, popular medicine and social geography. However, because those areas are traditionally only examined within a single disciplinary tradition (history of medicine, art history, folklore, literature, etc.), contradictions and incompatibilities between different forms of understanding the criminal body do not necessarily present themselves. Our way of working has demanded that we address incommensurable beliefs and the significance of context.

- We take as our starting point the idea that the criminal corpse, even when life had left the body, was still a powerful object. It had social and symbolic power, which was manipulated by the State and other interest groups, as well as medicinal and curative power.
- It was a key source for the creation and deployment of the power of scientific and medical knowledge, and of discretionary judicial power. In some cases, it was also understood to have its own agentive power.
- The power of the criminal corpse was harnessed in the promotion of particular social interests (in relation to class, gender and race for example). It was used instrumentally in the construction of knowledge and in medicine.
- Beliefs about, and the idea of, the criminal corpse informed popular culture and the development of historically situated normative ethics which continue to affect our beliefs about the dead body today.
- Culturally the criminal body is located at the overlap of several different traditions of discourse and practice and is a lens through which tensions such as normal/abnormal, and ethical/unethical can be explored in historical context.

In the overlap between modes of knowledge about the criminal body we find not only shared concerns and 'leakage' of assumptions and values from one area to another, but also contradictions and incommensurabilities which challenge us to develop more sophisticated understandings about how knowledge, belief, practice and resistance were produced and reproduced in the past.

Part 1 of this book has set out its aims, and outlined the historical lineage of the post-mortem punishments of the Murder Act. Part 2 focuses our examination of the criminal corpse during that period and looks in more detail at the tensions between retribution and humanity, deterrence and justice, science and prurience that characterise it.

NOTES

1. See, Dyer, C. (2005), *An Age of Transition? Economy and Society in England in the Later Middles Ages* (Oxford: Clarendon Press).
2. Russell, K.F. (1987), *British Anatomy 1525–1800: A Bibliography of Works Published in Britain, America and on the Continent* (London: St Pauls Biographies), p. xviii.

3. Ibid., p. xxi.
4. See, Park, K. (1995), 'The Life of the Corpse: Division and Dissection in Late Medieval Europe', *Journal of the History of Medicine and Allied Sciences*, Vol. 50, Issue 1, 111–132.
5. Ibid., p. 119.
6. See, Sawday, J. (1995), *The Body Emblazoned: Dissection and the Human Body in Renaissance Culture* (Abingdon: Routledge), p. 55.
7. See, Poole, S. (2015), '"For the Benefit of Example": Crime-Scene Executions in England, 1720–1830', in Ward, R. (ed.), *A Global History of Execution and the Criminal Corpse* (Basingstoke: Palgrave), pp. 71–101.
8. See, Davies, O. and Matteoni, F. (2016), *Executing Magic: The Power of Criminal Bodies* (Basingstoke: Palgrave).
9. See, Harland, J. and Wilkinson, T. (1882), *Lancashire Folk-Lore* (Manchester: John Heywood), p. 128; Weever, J. (1631), *Ancient and Funerall Monuments with in the United Monarchie of Great Britaine, Ireland and the Islands Adjacent* (London: Thomas Harper), p. 45.
10. See, Foucault (1978), *A History of Sexuality, Vol I: The Will to Knowledge* (Harmondsworth: Penguin).
11. See, Harris, O.J.T., Robb, J., and Tarlow, S. (2013), 'The Body in the Age of Knowledge', in Robb, J. and Harris, O.J.T. (eds.), *The Body in History* (Cambridge: Cambridge University Press), pp. 164–195, quote on p. 172.
12. See, Lewis, C.S. (1964), *The Discarded Image* (Cambridge: Cambridge University Press), p. 148.
13. See, Werness, H.C. (2006), *Continuum Encyclopedia of Animal Symbolism in World Art* (New York: Continuum), p. 37.
14. See, Aziz et al. (2002), 'The Human Cadaver in the Age of Biomedical Informatics', *Anatomical Record*, Vol. 269, Issue 1, 20–32.
15. See, Ghosh, S.K. (2015), 'Human Cadaveric Dissection: A Historical Account from Ancient Greece to the Modern Era', *Anatomy and Cell Biology*, Vol. 48, Issue 3, 153–169.
16. Siraisi, N.G. (1990). *Medieval and Early Renaissance Medicine: An Introduction to Knowledge and Practice* (Chicago: University of Chicago Press).
17. See, Sawday, J. (1995), *The Body Emblazoned: Dissection and the Human Body in Renaissance Culture* (Abingdon: Routledge).
18. See for example, Thomas, J. (2004), *Archaeology and Modernity* (London: Routledge).
19. See, Sawday, J. (1995), *The Body Emblazoned: Dissection and the Human Body in Renaissance Culture* (Abingdon: Routledge).
20. See, Nixon, A. (1612), *The Dignity of Man, Both in the Perfections of His Soule and Bodie*, p. 8.

21. See, Tarlow, S. (2010), *Ritual, Belief and the Dead in Early Modern Britain and Ireland* (Cambridge: Cambridge University Press).
22. See, Hillman, D. and Mazzio, C. eds. (1997), *The Body in Parts. Fantasies of Corporeality in Early Modern Europe* (London: Routledge), quote at p. xi.
23. The British Library catalogue lists in the half century before 1625, among others, *The Anatomie of the Mind* (Thomas Rogers 1576), *The Anatomie of Wit* (John Lyly 1579), *The Anatomie of Abuse* (Philip Stubbes 1583), *The Anatomie of Fortune* (Robert Greene 1584), *The Anatomie of Absurdity* (Thomas Nash 1590), *The Anatomie of Popish Tyrannie* (Thomas Bell 1603), *The Anatomie of Sorcery* (James Mason 1612), *The Anatomie of Mortality* (Strode 1618), *Follie's Anatomie* (Henry Hutton 1619), *The Anatomie of Vanitie* (Richard Brathwait 1621), *The Anatomie of Conscience* (Immanuel Bourne 1623), *The Anatomie of Protestancie* (O.A. 1623) and *The Anatomie of the Roman Clergy* (George Lauder 1623). See, Tarlow, S. (2010), *Ritual, Belief and the Dead in Early Modern Britain and Ireland* (Cambridge: Cambridge University Press).
24. See, Tarlow, S. (2013), 'Cromwell and Plunkett: Two Early Modern Heads Called Oliver', in Kelly, J. and Lyones, M. (eds.), *Death and Dying in Ireland, Britain and Europe: Historical Perspectives* (Dublin: Irish Academic Press), pp. 59–76.
25. The origins of the use of a stylised Guy Fawkes mask in late twentieth-century protest derive from the graphic novel *V for Vendetta* (1989) written by Alan Moore and illustrated by David Lloyd, which formed the basis for the 2005 blockbuster film by the same name, directed by James McTeigue and written by The Wachowskis. On the use of Guy Fawkes imagery in anti-government protest, see Call, L. (2008), 'A Is for Anarchy, V. Is for Vendetta: Images of Guy Fawkes and the Creation of Postmodern Anarchism', *Anarchist Studies*, Vol. 16, Issue 2, 154–172.
26. See for example, Weber, M. (1904–1905/1970), *The Protestant Ethic and the Spirit of Capitalism* (London: Allen & Unwin, 1970 edition); Tawney, R.H. (1926), *Religion and the Rise of Capitalism* (London).
27. See, Ariès, P. (1981), *The Hour of Our Death* (Harmondsworth: Peregrine), quote at p. 298.
28. See, Gordon, A. and Rist, T. eds. (2013), *The Arts of Remembrance in Early Modern England: Memorial Cultures of the Post Reformation* (Farnham: Ashgate).
29. See, Brown, T. (1979), *The Fate of the Dead: A Study in Folk-Eschatology in the West Country After the Reformation* (Cambridge: D.S. Brewer), quote at p. 16.
30. See, Sherlock, W. (1690), *A Practical Discourse Concerning Death* (London: W. Rogers, 2nd edition), quote at p. 53.

31. See, Boyd, Z. (1629), *The Last Battell of the Soule in Death* (Edinburgh: Heires of Andro Hart), quote at p. 84.
32. See, Owen, K. (2000), *Identity, Commemoration and the Art of Dying Well: Exploring the Relationship Between the Ars Moriendi Tradition and the Material Culture of Death in Gloucestershire, c.1350–1700A.D.* (Oxford: British Archaeological Reports).
33. See, Houlbrooke R. (1999), 'The Age of Decency: 1660–1760', in Jupp, P.C. and Gittings, C. (eds.), *Death in England: An Illustrated History* (Manchester: Manchester University Press), pp. 174–201.
34. Ibid., p. 183.
35. Ibid., quote at pp. 184–185.
36. Henri Misson M., *Misson's Memoirs and Observations in His Travels over England, with Some Account of Scotland and Ireland. Written Originally in French and Translated by Mr Ozell* (London, 1719).
37. As cited in, Gittings, C. (1984), *Death, Burial and the Individual in Early Modern England* (London: Routledge), p. 64.
38. William Andrews in *Old Church Lore* (William Andrews & Co., The Hull Press; London, 1891; pp. 191–194) cites a number of ballads that make a joke of the condemned man choosing execution over marriage. Oloudah Equiano, in his *Interesting narrative*, reports witnessing a man saved from death in this way in New York in 1784: 'One day a malefactor was to be executed on a gallows; but with a condition that if any woman, having nothing on but her shift, married the man under the gallows, his life was to be saved. This extraordinary privilege was claimed; a woman presented herself; and the marriage ceremony was performed'.
39. See, Lake, P. (1993), 'Deeds Against Nature: Cheap Print, Protestantism and Murder in Early Seventeenth-Century England' in Sharpe, K. and Lake, P. (eds.), *Culture and Politics in Early Stuart England* (Stanford: Stanford University Press), pp. 257–284, esp. p. 274.
40. Ibid., quote at p. 276.
41. Ibid., p. 276.
42. See, Laqueur, T.W. (1989), 'Crowds, Carnival and the State in English Executions, 1604–1868', in Beier, A.L., Cannadine, D., and Rosenheim, J.M. (eds.), *The First Modern Society: Essays in English History in Honour of Lawrence Stone* (Cambridge: Cambridge University Press), pp. 305–355.
43. See, Beattie, J.M. (1986), *Crime and the Courts in England 1660–1800* (Oxford: Oxford University Press).
44. Quoted in, Linebaugh, P. (1975), 'The Tyburn Riot Against the Surgeons', in Hay, D., Linebaugh, P., Rule, J.G., Thompson, E.P., and Winslow, C. (eds.), *Albion's Fatal Tree* (New York: Pantheon Books), pp. 65–118, at p. 83.
45. See, Laqueur, T.W. (2015), *A Cultural History of Mortal Remains* (Woodstock: Princeton University Press), p. 316.

46. See, Gittings, C. (1984), *Death, Burial and the Individual in Early Modern England* (London: Routledge), p. 67.
47. Ibid., quote at p. 71.
48. See for example, Kilfeather, S. (2002), 'Oliver Plunkett's Head', *Textual Practice*, Vol. 16, Issue 2, 229–248 and Tarlow, S. (2008), 'The Extraordinary Story of Oliver Cromwell's Head', in Robb, J. and Borić, D. (eds.), *Past Bodies: Body-Centred Research in Archaeology* (Oxford: Oxbow Books), pp. 69–78; Tarlow, S. (2013), 'Cromwell and Plunkett: Two Early Modern Heads Called Oliver', in Kelly, J. and Lyones, M. (eds.), *Death and Dying in Ireland, Britain and Europe: Historical Perspectives* (Dublin: Irish Academic Press), pp. 59–76.
49. See, Bennett, R. (2017), '"A Candidate for Immortality": Martyrdom, Memory, and the Marquis of Montrose', in McCorristine, S. (ed.), *When Is Death? Interdisciplinary Perspectives on Death and Its Timing* (Palgrave Macmillan).
50. For example, Foucault, M. *Discipline and Punish* (Pantheon 1977), *The History of Sexuality* (1978, 1985, 1986). See, Elias, N. (1939/2000), *The Civilising Process*, trans. Edmund Jephcott (Oxford: Blackwell, revised edition); Weber, M. (1904–05/1970), *The Protestant Ethic and the Spirit of Capitalism* (London, Allen & Unwin, 1970 edition).
51. See for example, Linebaugh, P. (1991), *The London Hanged: Crime and Civil Society in the Eighteenth Century* (London: Verso), p. 42.
52. See, Bailey, V. (2000), 'The Death Penalty in British History', *Punishment and Society*, Vol. 2, Issue 1, 106–113.
53. See, Lake, P. (1993), 'Deeds Against Nature: Cheap Print, Protestantism and Murder in Early Seventeenth-Century England', in Sharpe, K. and Lake, P. (eds.), *Culture and Politics in Early Stuart England* (Stanford: Stanford University Press), pp. 257–284, quote at p. 274.
54. Ibid., p. 274.
55. See, Laqueur, T.W. (1989), 'Crowds, Carnival and the State in English Executions, 1604–1868', in Beier, A. L., Cannadine, D., and Rosenheim, J.M. (eds.), *The First Modern Society: Essays in English History in Honour of Lawrence Stone* (Cambridge: Cambridge University Press), pp. 305–355, quote at p. 327.
56. Ibid., p. 322.
57. See, Richardson, S. (1928), *Familiar Letters on Important Occasions* (London: Routledge, 1st edition, 1741), pp. 217–219.
58. Described in, Laqueur, T.W. (1989), 'Crowds, Carnival and the State in English Executions, 1604–1868', in Beier, A.L., Cannadine, D., and Rosenheim, J.M. (eds.), *The First Modern Society: Essays in English History in Honour of Lawrence Stone* (Cambridge: Cambridge University Press), pp. 305–355.

59. See, Howell, T.B. (1816), *A Complete Collection of State Trials*, Vol. II, Coll. 84 (London: Longman).
60. See, Gittings, C. (1984), *Death, Burial and the Individual in Early Modern England* (London: Routledge), pp. 70–71.
61. Discussed in, Kilfeather, S. (2002), 'Oliver Plunkett's head', *Textual Practice*, Vol. 16, Issue 2, 229–248; Tarlow, S. (2013), 'Cromwell and Plunkett: Two Early Modern Heads Called Oliver', in Kelly, J. and Lyones, M. (eds.), *Death and Dying in Ireland, Britain and Europe: Historical Perspectives* (Dublin: Irish Academic Press), pp. 59–76.

PART II

The World of the Murder Act

CHAPTER 4

Murder and the Law, 1752–1832

On 26 March 1752, the Act for 'better preventing the horrid crime of murder' passed into law after moving swiftly through the Commons and the Lords and receiving royal assent from George II. The establishment of the Murder Act, as it is known, was a significant moment in the history of British criminal justice. It stood without change or serious challenge for eighty years and is unique in British history. The Murder Act established systematic juridical procedures for the execution and, critically, the post-mortem punishment of convicted murderers.

From 1752 to 1832, the punishment for anyone convicted of murder, even members of the nobility,[1] was execution by hanging. There is little new in this, considering the eighteenth century was the time of the Bloody Code—the name given after the fact to Britain's eighteenth-century penal code because of the high number of capital crimes on the books: over 200 by the 1820s.[2] However, the sentence for murder did not end with the death of the condemned. The corpse of the convicted murderer was then sent for anatomisation and dissection or handed over to the sheriff to be hung in chains ('gibbeted'). In either case, the punishment inflicted on the criminal corpse was, by intention and in effect, highly visible.

Gibbeted bodies were suspended thirty feet in the air, and stayed in place for decades as they decayed on display. For those sentenced to anatomisation, crowds trooped through inns and other convenient sites to see bodies cut and spread open. Then corpses were carted off to the

© The Author(s) 2018
S. Tarlow and E. Battell Lowman, *Harnessing the Power of the Criminal Corpse*, Palgrave Historical Studies in the Criminal Corpse and its Afterlife, https://doi.org/10.1007/978-3-319-77908-9_4

much more secluded anatomy rooms where they would be dissected to their extremities, that is, until there was nearly nothing left, by surgeons or groups of medical trainees. These very public and intensely sensory punishments visited ignominy and humiliation on the condemned. They also deliberately prevented the interment of the body and performance of the customary religious and cultural rituals, denying both the condemned and their family the comfort and finality of a decent burial.

Reserved for the most unnatural of crimes, the Murder Act was intended to deter through terror the commission of future murders. In practice, the effects were both broader and more complex. To understand the changing power of the criminal corpse in Britain, the eighty-year lifespan of the Murder Act, defined by state-mandated and court-ordered punishments carried out on criminal corpses, is critical. This period witnessed a steep rise in the number of capital offences and a simultaneous and paradoxical overall reduction in the number of executions carried out, apart from a few moments in the 1750s, 1780s, and 1810s when execution numbers showed sudden and sharp spikes. Major developments in theories and practices of punishment associated with changing beliefs about the body took place in the eighteenth and early nineteenth centuries. In particular, older forms of punishment involving injury and maiming were gradually phased out and replaced by new types of punishment based on removal of the offender from society, imprisonment and hard labour.[3] The use of the pillory, the stocks, whipping, and burning gave way to incarceration in new facilities on land and on water, and transportation overseas served the dual purpose of punishment and extending the power and profitability of the Empire.

This chapter focuses on the law in order to trace the history of codified post-mortem punishment from its conception—by lawmakers, judges, executioners, the accused and others—to its creation under the Murder Act. We begin with an examination of the creation of the Act and the context in which it was proposed. Then we investigate how the Act operated in practice, including its uneven application across different geographical regions (the London metropolis, peripheral regions of England, Scotland, and colonial contexts such as Wales, Ireland and the overseas colonies). Next, we consider the impact of the Act, and the relationship of punishment under the Act to other forms of punishment in use during the same period for serious crime—in particular, transportation and incarceration on land and water—and the fate of the criminal corpses created by them. We conclude this chapter with attention to when and why the Murder Act was repealed, and its enduring legacy.

MAKING THE MURDER ACT

The Murder Act was created during a period of unprecedented attention to crime and punishment in England and in the midst of a press-fuelled moral panic about crime in the capital. In February 1751, a House of Commons committee was established to investigate felonies and other offences. This was the first time a committee was tasked with a specific focus on crime and punishment. It included all of the Members of Parliament for London, Middlesex and Surrey, giving the committee a distinctly metropolitan character, and by 1752 had produced three orders: the Disorderly Houses Act, the Confinement at Hard Labour Bill, and the establishment of the Pawnbrokers Committee. The activities of the Felonies Committee made no sign towards the terms that would appear in the same year in the Murder Act. Rather, the Commons ordered Sir William Yonge and Sir George Lyttleton to bring a bill 'for the better preventing the horrid crime of murder' on 10 February 1752.[4] Why was murder singled out for special attention in this way?

Historian Richard Ward has convincingly argued that the Murder Act was introduced in response to a short-lived panic that arose in London as a result of several homicides that took place in the capital during late 1751 and early 1752, which were covered extensively in the press.[5] In the decade preceding the panic, murder prosecutions had been at or below the annual average for 1720–1759, and if we consider only civil cases (excluding the Admiralty) the 'spike' in murder prosecutions in London in 1751 actually only consists of 2 capital convictions. The increase was not as dire as may have appeared, but this was certainly not how it was portrayed in the press. The *Daily Advertiser*, *Penny London Post*, and *Read's Weekly Journal* all wailed about the impending deluge of prosecutions, and gave unprecedented attention in their pages to the crimes and prosecutions of murderers.[6] Crimes certainly occurred, but it was press reporting on (or more accurately, sensationalising) such crimes that caused murder to become 'a problem greater than the sum of its parts.'[7] Nonetheless, the press-induced panic in the mid-eighteenth century spurred rapid action from government in no small part because the primary audience for printed crime literature, British society's middling ranks, included some of the key decision makers for the criminal justice system—particularly those in London.[8]

It took less than a month to prepare the bill and present it before Parliament. After some discussion in the Commons during the first three weeks of March 1752, the bill was passed and put before the Lords who made one significant amendment that stipulated severe punishment for anyone attempting to interfere in the post-mortem punishments mandated under the Act. On 26 March, the Act became law and stood until 1832, when key clauses began to be repealed.

The Act hastened sentencing and execution for those convicted of murder but most strikingly, under the Act convicted murderers were denied the comfort of the prospect of a decent burial; instead their sentence would include the terror of knowing their body was destined for public anatomisation and dissection or being hung in chains on public display and left to rot. The Act also changed the timeline of punishment for murder, making it more difficult for some to petition for pardon. Over its eighty-year lifespan, the impacts and entanglements of the Act went far beyond its original intent with significant implications for the developing professionalisation of medicine, but the Murder Act was first and foremost a tool for social control designed to create and harness terror to punish and deter.

The Murder Act introduced five specific measures designed to increase the terror of punishment and viscerally distinguish the crime of murder from all others. The clause stipulated 'the sentence of death should henceforth be passed upon murderers in open court immediately after conviction'. This shortened the possible window during which appeals or bargaining could occur that might soften the anticipated sentence. Similarly, the second clause required that the execution of those condemned under the Act be carried out two days after sentencing, with the exception that if it should fall on a Sunday the execution would take place on the following day (Monday). The combination of the first two clauses meant that punishment would now be much more rapidly inflicted. The third clause directed those convicted of murder to be held in solitary confinement and allowed a diet of only bread and water. The conduct of those condemned to die following capital conviction in Britain in the eighteenth century who had the means to pay for it can be described as riotous. It was not unusual for individuals in this situation to pay to have visitors and prostitutes brought to see them, and to entertain in as lavish a fashion as was possible. The well-off spent money on food, alcohol, and company to make the most of their last days, and those who lacked the means might be able to arrange sale of their corpse to anatomists and use the coin to procure entertainment or a measure of respite from the prospect of the punishment to come. Mandating solitary confinement, except of course for visits from clergy to promote proper contrition and religious submission, and such a spare diet made it much more difficult for the condemned to ignore or subvert the impact and lesson of the punishment that awaited them.

The fourth clause of the Act is the one that sets it apart from all other penal legislation in British history: after execution by hanging, the bodies of convicted murderers were not to be allowed burial until they had been either anatomised and dissected or hung in chains. The use of dissection and hanging in chains—known as 'gibbeting'—were not new developments. Some individuals convicted of capital crimes before 1752 were gibbeted as an additional punishment and object lesson, and the corpses of some executed criminals were given over to the medical men through royal and other grants.[9] The power to punish the criminal corpse was based on an understanding common in England since the fourteenth century, but never enshrined in law, that 'the bodies of executed felons were at the disposal of the king.'[10] This was also the basis of earlier allowances by royal decree of a small fixed number of criminal corpses to the College of Barber Surgeons and College of Physicians for study (see Chapter 3). Before 1752, post-mortem punishments were discretionary in all but the crime of treason, for which sentence to being hanged, drawn, and quartered involved post-mortem punishment and public display of the body. The Murder Act formalised existing practices rather than developing new post-mortem punishments. In so doing, it established post-mortem punishment as both legally mandated *and* systematic for the first time in Britain.[11]

Concern over the potentially riotous antics of those awaiting execution, addressed by the first three clauses, was dwarfed by concerns linked to the behaviour of the disorderly, boisterous, and rowdy audiences that attended executions in eighteenth-century Britain. The Murder Act's final clause sought to prevent interference with the punishment of convicted murderers: attempting to rescue the condemned before execution was made punishable by death. The Act also made any attempt to rescue the murderer's corpse punishable by transportation overseas. It was designed to ensure that the punishments to which murderers were sentenced had as little chance as possible of being interrupted by the public or the family and friends of the condemned. Peter Linebaugh has demonstrated that the surgeons and their representatives who claimed bodies from the gallows—whether through legal means, such as the corpses accorded them through royal grant and upheld by parliament, or through only semi-legal means, such as the prearranged purchase of a corpse from a condemned person or from the hangman—were the most common targets of disorder at the foot of the gallows before the advent of the Murder Act.[12] The preventative clause in the Murder Act sought to reduce the danger both of riot or unrest that might cause injury and property damage, and avoid subversion of the full punishment to which a murderer had been sentenced.

The motivations behind four of the five key clauses of the Murder Act seem clear, considered in the context in which the Act was created. However, it is less obvious why parliament should consider harm inflicted on a criminal's corpse to be either useful or appropriate to the goals of maintaining social cohesion or state control. Historian Peter King has discovered that there were debates before the advent of the Murder Act about how to most effectively add to the terror and infamy of the punishment meted out against murderers. These discussions were based on a perceived need to set murder apart from other crimes. When the eighteenth century began there were over 60 capital crimes, and by the end of the century the number had grown to over 200.[13] Penalising small theft *and* murder with the same punishment—death by hanging—seemed to offer little deterrent to, or differentiation from, the commission of one of the most violent and socially transgressive crimes. Further, as Elizabeth Hurren has identified, the principle of *lex talionis*—the idea that punishment should correspond in both degree and kind to harm done by the wrongdoer—directed that some additional punishment was required to restore social balance, and the impact of murder, not just on the victim but on society and social cohesion, meant that some additional punishment was needed.[14] The admittedly brief mid-century panic about violent murder in London spurred the creation of the Murder Act, but it was the issues of deterrence, differentiation, retribution and rebalance that gave rise to its specific content.

But harming criminal corpses as an additional punishment for murder was not the only option considered in the mid-eighteenth century. The possibility of using aggravated punishment, that is, pre-execution forms of physical retribution for the wrongdoing, was also proposed. In other parts of Europe, forms of aggravated punishment such as breaking on the wheel were used into the 1830s to punish individuals convicted of particularly heinous crimes.[15] The prospect of such protracted and excruciatingly painful torture and the humiliation of suffering it in public would no doubt have added the 'further mark of terror and infamy' intended by the Murder Act. It has proven difficult to unravel the precise reasons behind the creation of the specific content of the Murder Act and the punishments it stipulated, due to a lack of parliamentary reporting.[16] What we do know is that up to a fairly late stage, additional punishments both pre-execution and post-execution were debated in the press and in parliament.

The option eventually taken up as the way to 'add some further terror and peculiar mark of infamy to the punishment of death'[17] was to inflict humiliating punishments on the body of the convicted murderer after their execution. The condemned would not suffer physical pain if their corpse were harmed. So how was harming a dead body an additional punishment?

The terror of post-mortem punishment arose from common and strongly held concerns regarding bodily integrity and proper burial.[18] The exposure of the body was a source of humiliation to the condemned and shame to their family and friends. Further, the desecration and destruction of the body precluded the anticipatory comfort of a burial in accordance with one's faith. Ward has noted that the formal use of post-mortem punishment and the prevention of the burial rites associated with Christian salvation (the 'proper' burial of an intact body) represented an attempt to assert the authority of the law over that of God by placing 'decisions over the spiritual salvation over criminals within the hands of the secular courts.'[19] Although there is nothing in Christian doctrine that requires the burial of a whole body for resurrection (and indeed, St Augustine is quite clear that God would be able to assemble a body for resurrection even in the case of those who had been consumed by wild beasts, or burned in a fire), there does seem to have been a superstitious feeling that the lack of a 'decent' burial would have repercussions in the afterlife. The exposure of criminal (and criminalised) bodies and denial of burial and its associated rites and comforts permanently excluded the condemned from their community. The post-mortem harms detailed in the Murder Act were intended to punish murderers beyond their death, and to humiliate and disgrace them and their kin beyond the execution.

Post-mortem punishment under the Murder Act was a calculated response to the needs of mid-eighteenth-century Britain. It satisfied the principle of *lex talionis*, acted as a deterrent to the commission of future murders by providing an object lesson to the public, and inspired horror in the condemned. It was 'above all else designed to be terrifying, exemplary and shameful.'[20]

Hanging in chains had been in use as a discretionary punishment since at least the seventeenth century, and its use under the Murder Act followed the form practised before 1752 (details of gibbeting are taken up in Chapter 6). The punishment of 'anatomisation and dissection' however, deviated from the previous use of criminal corpses by medical men. Under the grants made to surgeons and anatomists by the Crown, criminal corpses were used for medical research and training purposes, but only infrequently in forms accessible to the public. The post-mortem punishment of anatomisation and dissection critically involved a very public element, during which the execution crowd could witness first-hand the cut corpse spread open to their view. This allowed the public to see justice done in that the full sentence had been executed and also to partake in the deterrent example intended by the spectacle (see Chapter 5). Both post-mortem punishments detailed in the Murder Act involved public exposure and desecration of the body, and resulted in the obliteration of the murderer's corpse.[21] But how did the Murder Act function in practice, and did these additional punishments achieve their intended effects?

Making Criminal Corpses

Criminals do not die by the hands of the Law. They die by the hands of other men.[22]

Over its eighty-year life, 1166 individuals were convicted of murder and sentenced under the Murder Act. Of these, 80% were sentenced to anatomisation and dissection and their corpses were handed over to the medical men following execution by hanging. A significantly smaller proportion, only 12%, was sentenced to hanging in chains, and the corpses of these individuals were gibbeted following execution on the gallows. Though the chances of a pardon for capital crimes other than murder fluctuated, they were consistently high for non-murder capital crimes—often hovering around 75%—the likelihood of securing a pardon for the convicted murderer was far more remote. Only 8% of those convicted under the Murder Act received a pardon and escaped execution and post-mortem punishment.[23] Here we trace the process of producing criminal corpses from individuals accused of murder, including sentencing, pardoning and execution. We also consider the outcome of a strategy attempted by a few individuals convicted and sentenced under the Act to avoid their grisly fates. Finally, we survey the uneven geographic application of the Murder Act across Britain. This approach provides insight into how the Act operated in practice, not just in principle. Chapters 5 and 6 take up the story of the criminal corpses produced under the Murder Act, and follow them into the spaces, experiences, and impacts of anatomisation and dissection and of hanging in chains, respectively.

In eighteenth- and nineteenth-century Britain, serious crimes, including murder, were tried by Crown courts. Outside London, such trials occurred at the assizes, which took place twice a year in large towns according to a regular annual schedule. In the capital, such cases were tried at the Old Bailey. Following trial, those convicted of murder received their sentence in open court immediately following conviction. The judge donned a small black cap (on top of the required powdered wig) and passed sentence of death. Under the Act, the judge was also responsible for sentencing the condemned to the post-mortem punishment of anatomisation and dissection. Should a judge deem gibbeting to be a more appropriate punishment, they then gave the order for the murderer to be hung in chains which replaced the initial sentence of anatomisation and dissection.

The first conviction for murder under the Act occurred in June 1752 in London. Thomas Willford had been indicted for the murder of his wife, Sarah (née Williams), in May of the same year. Sarah had been brutally attacked with a knife on 25 May, and nearly decapitated. The record of the trial is short, and there seems to have been little question of Thomas's guilt—indeed, he offered no evidence in his own defence. The Ordinary of Newgate's printed account of executed Old Bailey offenders, dated 2 July 1752 noted that this was the 'first case after the new act of parliament', and recorded the sentence passed by the judge: 'Guilty Death. He received sentence immediately to be executed on the Thursday following, (being cast on the Tuesday before) and his body to be dissected and anatomized.'[24] Early and scrupulous attention to the clauses of the Act is evident: sentence was passed immediately after conviction, the date of execution was set for two days after sentencing, and the judge clearly indicated the post-mortem punishment chosen for the condemned.

After sentencing, there was still the possibility that a convicted murderer could be pardoned. However, judges were far less likely to show leniency to convicted murderers than those found guilty of property offences. By reducing the delay between conviction and sentencing, and between sentencing and execution, the Murder Act significantly reduced the time and therefore the opportunity to organise petitions and make convincing pleas for clemency. But this was not the case in Scotland, where the stipulations of the 1725 Disarming Act (11 Geo I c.26) mandated delays between sentencing and execution—not less than 30 days if the sentence was pronounced south of the River Forth, and 40 days if it was pronounced north of the Forth. The Murder Act did not repeal this clause, and as a result, those convicted of murder in Scotland had as much time as those convicted of other capital crimes to send petitions to London asking for the Royal mercy.[25]

It is important to mention the difference between the pardoning process in the metropolis (London) and the rest of the country. Key decisions, including the granting of a pardon, were made by the assize judges in the provinces. In the capital, the Recorder of London provided reports to the king and his cabinet on capital convictions. The committee to which the Recorder reported could grant a royal pardon to the condemned or a conditional pardon that commuted their punishment from death to other punishments such as transportation, hard labour, or penal servitude.[26] In the case of clearly proven homicide, however, the pardoning system was seldom applied.[27] Murder was considered morally execrable and contravened the biblical edict 'thou shall not kill'. Pardoning in these cases was not popular: there was a strong public desire to see justice done on the body of the condemned, and this retaliation was sanctioned in both popular imagination and common law.[28] To give an idea of how pardoning functioned with relation to convictions for murder in practice, in London, of the 170 people convicted of murder and sentenced to dissection and anatomisation between 1752 and 1832, only 12 people were pardoned (and 10 hung in chains).

The clauses of the Murder Act also left little opportunity for those involved in deciding and carrying out a sentence to exercise discretion. For judges, gaolers, and the sheriffs and medical men involved in carrying out punishment, deviating from the Act's five key clauses was difficult. Given the swiftness of sentencing after conviction, the short timeline between sentencing and execution, and the binary choice faced by the judge between dissection and gibbeting, there was little opportunity to adapt, evade, or soften the sentence in response to local or contextual factors. The strict terms of the Act and harsh punishments mandated for anyone interfering in the execution of a sentence of murder in this period go a long way to explain the paucity of clear instances of resistance. Beyond opportunity and deterrence, interfering in the punishments prescribed for convicted murderers was socially suspect.

The only sanctioned method of execution under the Murder Act was that the condemned be 'hanged by the neck until dead'. Execution was also governed by formal and informal protocols. Public execution sites regularly drew thousands—whether spectators or witnesses. To the dismay of the authorities, in the eighteenth century executions were rarely solemn occasions, and were instead treated by the public as capital entertainment events. Before the advent of the Murder Act, 'hanging days' at Tyburn usually involved the execution of groups of people, not single individuals, which helped give rise to large and riotous crowds eager to see the action. After 1752, the number of 'hanging days' increased due to the requirement that convicted murderers be executed two days after sentencing, or three should the second day fall on a Sunday.

Until 1783, those destined for execution in London travelled by open cart from Newgate prison to the gallows at Tyburn. The journey was a chance for the condemned to win the approval of the crowd by dying 'game', showing fortitude of spirit and disdain for the solemnity of the occasion. It was also a chance to drink at stops along the way, which provided some last enjoyment, could contribute to showing casual disdain for the event to come, and allowed many to anaesthetise themselves with alcohol before mounting the scaffold where they were met by the hangman and the priest. Close by, the crowd watched, bawled and conducted business, both legitimate and illicit (Fig. 4.1).[29] The beadles employed by the surgeons were also be on hand to collect corpses destined for anatomisation and dissection, or if gibbeting was the allocated punishment, the sheriff or his representative attended to collect the body.

Outside London, executions were also public affairs, and were conducted either at customary hanging sites or, as Steve Poole has shown, at sites directly connected with the crime of the condemned. These in situ executions drew even stronger connections between the punishment of the condemned and retribution or restitution for their crime than other public executions. They contributed to the symbolic rehabilitation of space made dysfunctional by particularly socially transgressive crimes.[30] These crime scene hangings also included the procession of the condemned from the gaol where they were held to the execution site. Between 1720 and 1830, 'at least 211 people were taken in procession

Fig. 4.1 The idle prentice executed at Tyburn. William Hogarth 1795 (Wellcome Collection)

to the scene of their crime to be hanged' in England.[31] However, there were increased costs associated with hanging outside customary sites, and this contributed to their noted irregularity in our period of study.[32] Whether the execution occurred at a place relating to a particular crime or in a customary spot, it was intended to serve as an example to the wider community, to increase the terror of the punishment, and to confirm to the public that justice had been done.

But public executions in general became increasingly expensive and unmanageable over the course of the eighteenth century. The cost of hiring men to provide additional security to prevent injury or property damage should the crowd become disorderly was significant. The size of the crowd could also cause problems when the large groups attracted by the execution spectacle snarled up roads for a long time and impeded commerce. Important, too, was that the 'carnivalesque' nature of the execution crowd demonstrated that the sombre lesson intended by public punishment was not necessarily the message the crowd took away from attending hangings.[33] Moreover, in London the site of the Tyburn gallows became highly desirable real estate in the second half of the eighteenth century, and the prospective (and lucrative) uses to which the land could be put were not compatible with the macabre and crowded execution days.

In 1783, Tyburn was abandoned as the site of execution in favour of an execution ceremony conducted on a platform attached to the exterior of Newgate prison. This ended the execution procession as the condemned now only had to travel from the prison to its wall. It also restricted crowd attendance and involvement because the space in which people could congregate to observe the event was much more limited. This reduced the scale of and opportunity for behaviours which did not accord with the intended moral lesson of the spectacle. The shift from public towards private capital punishment has been understood as relating to shifting sensibilities that found the gruesome spectacle distasteful and out of step with the changing social mores of the time, but Simon Devereaux has identified the move from Tyburn to Newgate not so much as 'a departure towards more modern practices' but as 'one of the last stages of substantial innovation in an older system of thinking' that represented an effort to preserve by improving upon a 'still repugnant' practice.[34] We return to the process and moment of execution in Chapter 5 including the way death occurred on the gallows in eighteenth- and nineteenth-century Britain. We take up the story and challenges of the execution crowd and their movement from places of execution to places of post-mortem punishment in Chapters 5 and 6.

Peter King and Richard Ward have identified striking variations in how the Bloody Code was applied in different parts of the United Kingdom. In particular, the greater the distance from the centre (London), the more limited the political reach and power of the British state. The result was that in peripheral regions of the country—and strikingly in regions where Celtic language traditions survived—it was much less likely that crimes which by statute should be punished with death would actually result in execution.[35] This phenomenon was particularly visible in a strong reluctance to hang property offenders. King and Ward noted, 'If the Bloody Code was often a dead letter on the periphery, it was primarily because the citizens of those areas chose to make it so.'[36]

But did this trend hold beyond property crimes punishable by death into other capital crimes? In a word, no. Murder was widely understood and felt to be a very different kind of crime than offences against property. The outer counties of England and Wales did not hesitate to employ the full weight of the law against murderers. The closer social cohesion between citizens in peripheral regions that arose as a result of more flattened social strata outside the capital and a higher degree of interaction and interdependence occasioned by rural life may have worked to reduce the number of individuals hanged under the Bloody Code, but violent and deeply socially transgressive crimes such as murder were not regarded any more indulgently. In the western and northern counties of Wales and England, hangings of murderers occurred at a greater relative rate to hangings for property offences than in the metropolis.[37] Though the number of murderers was far higher in the capital compared to any other region, the Murder Act was applied more uniformly across the United Kingdom than other elements of the Bloody Code.

IMPACTS, INTENDED AND OTHERWISE, OF THE MURDER ACT

Did the Murder Act achieve its intended ends, and what impacts, intended and unintended, did it have? To answer these questions, we consider first whether the Act was successful in adding a 'further mark of terror and infamy' to the crime of murder and do so by investigating the most extreme actions taken by those condemned under the Murder Act to avoid the punishment that awaited them. Then we investigate whether the Act deterred the commission of future murders and functioned as an effective social control. Finally, we examine punishment in the context of British crime and justice during the life of the Act (1752–1832) to find out how the provisions of the Murder Act in both law and practice fit into discourses of punishment more broadly in Britain, and with what import for the intended impacts of the Act.

An individual convicted of murder in Britain between 1752 and 1832 was not only sentenced to death by hanging and to post-mortem punishment, but was also subject to a legal requirement that these punishments would take place quickly and under sombre conditions. Visits to murderers awaiting execution were limited to the clergy. Along with the brevity and austerity of the pre-execution interval, this was intended to provoke an appropriately penitent spirit in the condemned. Confession and repentance continued to be the desired goal, as had been the case for some centuries already (see Chapter 3). It is difficult to determine to what extent the stipulations of the Act helped to achieve confessions and repentance of the condemned as the recorded versions of these are neither more frequent nor more heartfelt than in previous periods. The pamphlets that frequently offered supposedly 'true' accounts of such lamentations were often written in advance so these popular accounts could be sold at the executions, making them unreliable sources for our purposes. One way of more effectively gauging the impact of the provisions of the Murder Act is to consider the instances in which condemned murderers in our period of study took extreme measures to escape elements of their punishment.

In six cases, individuals committed suicide after between being sentenced under the Murder Act and being taken to their execution. This represents a tiny fraction—only 0.5% of those convicted and punished under the Act—but their stories are significant for what they reveal about the lengths to which people were willing to go to avoid elements of the punishments that awaited them.

In the early modern period, self-killing was considered a 'desperate sin' in the eyes of the church and a felony under the law,[38] and the bodies of suicides were fiercely contested objects.[39] Anyone who willingly took their own life was at risk of being found culpable of committing a crime. A coroner's inquest could determine that a suicide was guilty of felo de se—committing a felony against themselves. The punishment for this crime was that all possessions were forfeit to the Crown and the body denied decent Christian burial. In much of England the body was interred at a crossroads or in a public way, face down, and depending on local custom, a wooden stake might be driven through the body.[40] This post-mortem punishment was intended in part to protect the community: the souls of suicides were known to be restless and malevolent. Staking anchored the ghost, and burying the corpse at a crossroads confounded the revenant, ensuring that it would not be able to return to the community and inflict harm on the living.[41] However, the legal penalty (confiscation of goods) and the customary penalty (post-mortem punishment) were only performed in cases of adjudged felo de se. Suicides could also be found non compos mentis and as such, not guilty of self-murder by reason of insanity or disordered mind.

By the second half of the eighteenth century, beliefs about suicide had changed dramatically. Juries were less and less inclined to render verdicts of felo de se, and instead in the vast majority of cases found suicides non compos mentis. This shift was due to a combination of factors. Most significant were a reluctance to impoverish and therefore punish the families of individuals who committed suicide and the decline of the power of religion and folklore to explain suicide in favour of medical explanations.[42] Where a finding of non compos mentis was returned, goods were not forfeit and the deceased was allowed a quiet, night-time, Christian burial, often in the north side of the churchyard.[43] Michael MacDonald identified a striking drop in the proportion of verdicts of felo de se in cases of suicide from the mid- to the late eighteenth century.[44] This trend continued into the nineteenth century; however, there was still a minority of cases in which felo de se was consistently returned. These fell into two groups. The first was made up of marginal community members, including outsiders, paupers and those in disgrace. These individuals had little in the way of value or community ties to the places where they died, and juries were less likely to exercise discretion on their behalf.[45] If they had goods, these were forfeited, and the bodies buried in public highways. The second group was criminals. In fact, from about 1760, the felo de se verdict was principally used as a way to punish individuals indicted or convicted of a crime who would otherwise 'escape' punishment by taking their own lives.[46]

The circumstances of six individuals whose death by suicide prevented their execution under the Murder Act—Francis David Stirn of London (d. 1760), John Bolton of Yorkshire (d. 1775), Joseph Armstrong of Gloucestershire (d. 1777), John Fearon of Cumberland (d. 1791), William Birch of Gloucestershire (d. 1791) and Thomas Smith of Dorset (1804)—offer clear indications that they sought to escape the public spectacle of execution and bodily desecration of post-mortem punishment. In two of these cases, details allow a closer examination of the motivations of the condemned.

First, the case of Francis David Stirn, who was convicted of murdering Richard Matthews by gunshot on 12 September 1760. The Ordinary's Account suggests Stirn attempted to feign madness to escape conviction, and at his trial spoke at length about his wish and plans for self-destruction.[47] He was sentenced to execution at Tyburn in London after which his body was to be anatomised and dissected. At his sentencing, he asked for the use of a coach instead of the usual open cart for the journey to the gallows but it was denied, as 'it was the intention of the Legislature that such Criminals should be exposed to public View as a Terror to all Persons that they should not be guilty of the horrid Crime of Murder'. Hearing this, Stirn drank something from a pint pot, and fell ill later that evening and died about eleven o'clock that night.[48] Stirn committed suicide to escape the ignominies of the punishment that awaited him, both the public and protracted spectacle of the execution. As we understand from the mention of his belief in bodily resurrection in the Ordinary's Report, he was particularly motivated to avoid the public and then private cutting and despoiling of his corpse.

Joseph Armstrong was convicted for petty treason on 12 March 1777 for the murder of his master's wife. He was sentenced to be drawn to the gallows on a hurdle, hanged until dead, then dissected and anatomised.[49] At seven o'clock on the morning of his execution, Armstrong, who continued to maintain his innocence, asked the gaolers for a few minutes by himself to devote to prayer. In that time he 'took a little strap, which it is imagined his mother hid in the straw, and tying this round his neck, he fastened it to a nail in the wall, and then by a sudden jerk dislocated his neck, and died before the people could open the door.'[50] Armstrong seems to have waited until the last moment for the possibility of a reprieve in the form of a pardon. When that hope was exhausted and possibly with the help of his mother, Armstrong took his own life. He had a clear and strong desire to avoid the punishments that awaited him as a convicted murderer. Perhaps he was also trying to avoid being confirmed as a murderer by accepting or experiencing the punishments mandated for that crime.

Suicide has been called 'the most private... of human acts'.[51] Certainly, self-inflicted death represented a way to avoid the horrors of public execution including of the public procession to the gallows, the drama staged there, and the humiliation of having the visceral, vulnerable, and vicious death of the short drop witnessed by the huge and carnivalesque crowd. However, the self-killing of those accused of or condemned for murder was considered one of the 'most heinous forms of premeditated suicide.'[52] So, did these individuals succeed in avoiding the post-mortem punishments they feared?

Stirn was tried posthumously, found guilty of felo de se, and his body sent for dissection. Bolton's corpse was also sent to the surgeons for dissection. Armstrong's corpse was hung in chains in the neighbourhood of Cheltenham on the direction of one of the judges who had convicted him, in part to prove beyond doubt he was dead even though the execution had not taken place.[53] Fearon had been sentenced to anatomisation and dissection, but after killing himself the night before his execution was convicted of felo de se and his corpse sentenced to be buried at a crossroads with a stake through the body.[54] Birch's body was also buried at a crossroads.[55] Finally Smith, though sentenced to anatomisation and dissection, had his corpse hung from the gallows that was to have been his site of execution to 'gratify the curiosity of hundreds of spectators.'[56]

The treatment of these corpses shows that the state was invested in carrying out post-mortem punishment on the bodies of suicides who were convicted criminals. This reasserted authority in the face of the condemned's attempt to circumvent the public and post-mortem aspects of punishment by controlling the manner and timing of their own death. It also gave proof to the people that justice had been done, and demonstrated that those who broke the law had not escaped punishment by ending their own lives. The nature of the post-mortem punishment may have changed from sentencing, but in all six cases, some form of post-mortem punishment was carried out to affirm and demonstrate the power of the state and the criminal justice system.[57]

That these condemned murderers committed suicide did not change the fact that within three days of sentencing, they were dead and their corpses subject to punishment. So what did their actions accomplish? These men took back a measure of control at a time when agency was otherwise denied them. They succeeded in avoiding at least part of the punishment that awaited them, in particular their participation in the state-directed and crowd-mediated spectacle of execution. In some cases, they sought to avoid the post-mortem punishment to which they had been sentenced and were successful—but only in so far as a different post-mortem punishment was eventually carried out. That this group of men took such drastic action to escape even part of the punishment to which they had been sentenced is powerful testimony to the terror inspired by the provisions of the Murder Act. A similar observation was made in the late eighteenth century by Commodore Edward Thompson who noted that those punished under the Murder Act 'always confessed more dread at the dissection of their dead bodies than any particular distress about the death on the gallows.'[58]

In at least some cases, the Act successfully made the punishment for murder more terrifying and horrible for those convicted of the crime. But did the Act deter the commission of future murders? About thirty years after the introduction of the Act, Lord Loughborough remarked that the horror that came over condemned murderers when they were informed their bodies would not be buried but would be destroyed either on the gibbet or the slab made a strong impression on witnesses.[59] Of course, it is difficult to assess the deterrent effect of this law because of the challenge of finding traces of crimes that were not committed.

The possibility of obtaining a pardon has been identified as reducing the deterrent effect of the Murder Act. Ward has found that during this period, it was 'believed that pardons ultimately brought more men to the gallows than they saved from it'[60] because 'the royal pardons all-too-frequently granted to condemned malefactors only served to undermine the terror of the gallows.'[61]

Punishment in Britain changed significantly over the life of the Murder Act. Physical punishments such as branding or burning were phased out at the same time as confinement-based punishments were on the rise. Though death on the gallows, with or without post-mortem despoliation of the corpse, was extreme and final, it coexisted alongside other horrific punishments that also produced various types of 'death' of the condemned. The forced relocation of condemned criminals to overseas territories often with a prohibition against return—known as transportation—was a key punishment employed by the British government. Clare Anderson has argued that execution should be considered alongside transportation, which was often an alternative to the gallows, and which produced the social death of the condemned even if they survived the journey, labour, and living conditions to which they were subjected.[62] In instances where execution as an event might cause unrest, or fail to serve the intended aims of the criminal justice system, transportation and the horror of separation and the unknown that it entailed, provided a useful alternative. Like the post-mortem punishments in the Murder Act, transportation allowed the state to make productive use of the bodies of convicted criminals. Transported convicts were used to build the overseas infrastructure of empire, their labour expanding the power and reach of the state just as the gibbeted or dissected corpse reinforced domestic public awareness of state power.

UNMAKING THE MURDER ACT

Debates and discussions over the intent and nature of punishment in the eighteenth and nineteenth centuries were energetic and took place in legislative, judicial, public, and domestic spaces. Common themes in these debates include improving deterrence and terror in the punishments for murder by adding aggravated punishment—such as breaking on the wheel or other pain-based pre-execution punishments—or extending the post-mortem punishments of the Murder Act to other capital crimes. Significantly, the most heated exchanges about crime and punishment took place around the moral panics about crime that the eighteenth- and nineteenth-century press both reported and helped create.[63]

In the late eighteenth century there were two legislative efforts to extend the post-mortem punishments at the heart of the Murder Act to other capital crimes. William Wilberforce introduced the 1786 Dissection of Convicts Bill before the House of Commons in the wake of the war with America when a crime wave seemed to sweep the nation. It called for the bodies of executed criminals condemned for crimes including high treason, rape, arson, burglary, and highway robbery to be handed over for dissection.[64] The possibility of deterring crime through the terror of dissection was certainly a good fit as a response to the moral panic in London about the rising rates of violent crime. Though the Dissection of Convicts Bill was introduced by Wilberforce, it in fact originated with his close friend and medical advisor, William Hey, and was motivated by the insufficient legal supply of bodies to the anatomists. It was calculated that the Bill would have made available through legal means, on average, 70 additional bodies for dissection each year (an increase almost equal to the average annual number of corpses dissected under the Murder Act), and subjected those 70 'to the shame, ignominy and horror of anatomization.'[65] Though construed as mutually beneficial to criminal justice and medicine, the needs of criminal justice won over the needs of medical training and research. The 1786 Bill passed through the House of Commons but was later thrown out by the Lords: 'Applying dissection to offences beyond murder would, it was argued, undermine its effectiveness as a penal measure.'[66]

The second of these ultimately unsuccessful attempts was proposed to the House of Commons in 1796 by Richard Jodrell in the form of the Motion for the Dissection of Robbers and Burglars. Jodrell's concern was based on what he perceived as a recent and alarming increase in robberies and burglaries and his belief that the dread of dissection would 'serve to check this wave of criminality.'[67] Jodrell's particular concern was the prevention of bodysnatching, a crime of which he had a particular horror.[68] In this period, the theft of bodies from graveyards was on the rise and was an unpleasant and illicit way to source corpses for the growing needs of anatomy training. Had it passed into law, Jodrell's motion to extend dissection to the corpses of executed robbers and burglars could have benefitted anatomists and their students by significantly increasing the number of corpses legally available to them.

It was Jodrell's intention that such an increase would correlate directly with a significant decrease in bodysnatching. The motion was intended to serve two ends: first, by creating a stronger deterrent, it aimed to reduce the number of robberies and burglaries committed; second, by making more bodies legally available to the medical men, it would preclude the crime of bodysnatching. It was the deterrent terror of dissection, paradoxically, that spiked the motion's chances of becoming law. MPs considered dissection as a penal measure effective in the prevention of murder and were loath to 'break down the barrier which nature had established between murder and other crimes.'[69] Because of the perceived risk that the extension of post-mortem punishment to other offences would diminish the deterrent effect of the Murder Act and lead to an increase in murders committed, the 1796 Motion for the Dissection of Robbers and Burglars failed.

For eighty years, the Murder Act remained the state-sanctioned source of bodies for medical training, research and demonstration. Further, while it subjected a small proportion of the population to post-mortem punishment, in practice it also functioned to restrict those punishments from being extended more broadly to those convicted of other capital crimes. And, although it was becoming increasingly apparent that there was a real and pressing need for improved safe and legal access to bodies for anatomical training and research, in the late eighteenth century it was clear that with regards to the dissection of convicted criminals, the priority was that it serve the interests of the criminal justice system, not medicine.

In the end, it was medicine, and the gruesome set of crimes associated with filling the supply gap in corpses required by anatomists and medical men in training, that caused the Murder Act to be superseded by new legislation in 1832. In the first three decades of the nineteenth century, bodysnatching was becoming alarmingly frequent, and the desecration of graves for this grisly purpose was the cause of widespread public disgust and fear. As the legal pathway to obtain bodies for teaching and research, the Act was also linked in effect if not intention to the insufficiency of bodies available and therefore to the illegal activities and trade that arose to supplement the supply of corpses to the medical men during this eighty-year period.

In the 1820s, campaigns by surgeons for access to alternative sources of bodies for anatomical research and training gathered strength. In 1825 surgeons petitioned the Home Office for the bodies of people who died in prisons, workhouses, or infirmaries. But a strong connection had developed between murderers and dissection that created durable prejudice against the use of the corpses of innocent people for anatomical work. Jeremy Bentham was a strong proponent of increasing the stream of bodies legally available for medical research and training, and worked to challenge the negative associations of dissection with criminality. In addition to drafting a 'Body Providing Bill' in 1826 that included a section repealing anatomisation and dissection in the Murder Act, he very publicly directed his body to be given for dissection after his death.[70]

It is also important to mention the effect that changing sensibilities in the nineteenth century had on the provisions for post-mortem punishment in the Act. From the late eighteenth century, capital punishment was increasingly becoming a private affair. In London, executions were relocated from Tyburn to Newgate Prison, and punishment was being shifted from a sensational public spectacle to a private, solemn, grim, and in some ways more terrifying event. Public opinion began to turn against post-mortem punishment and against gibbeting in particular. Very few instances of hanging in chains took place in the nineteenth century in Britain, and the practice was increasingly seen as a barbaric and disgusting display that did little except cause nuisance and revulsion for those who passed by. The last gibbet erected in Britain in 1832 was taken down after only three days because of the energetic protests it provoked from the local populace (see Chapter 6). In the 1820s, there were instances of inhabitants preventing a judge from gibbeting a murderer in Lincolnshire and older gibbets being disassembled and brought down. Gibbeting was 'effectively dead as a sentencing option by the mid-1810s'.[71] In this, Britain was broadly in step with developments elsewhere in Europe where post-mortem punishments, including display of criminal corpses (whole or in pieces) was abolished in the Netherlands, Prussia, the German states and Norway between 1795 and 1842.[72]

The Anatomy Act was brought before parliament in 1831 and became law in 1832. It made unclaimed bodies legally available to licensed anatomists for dissection and medical teaching and research. Specifically, the Anatomy Act allowed medical men access to the bodies of the poor—those who died in the workhouse or in prison.[73] It superseded part of the post-mortem provisions of the Murder Act, and from 1832 sentencing to anatomisation and dissection was not part of the punishment for murder in Britain. The Anatomy Act did not address hanging in chains, because that punishment had fallen so far out of favour in Britain that it had largely been abandoned. In light of the public outcry following the two gibbetings that took place in the summer following the Anatomy Act's passage into law in 1832, the hanging in chains option of the Murder Act was repealed in 1834.

During its life, the Murder Act forged a strong and enduring connection between murder and dissection in Britain. Further, the Act bucked broader trends towards incarceration and away from bodily punishment in the eighteenth and nineteenth centuries. The spectacular, visceral post-mortem punishments required under the Act that we explore in depth in the next two chapters stand out and against the civilising trajectory popularly proposed for this period. Anatomisation and dissection and hanging in chains may have added the intended mark of 'terror and infamy' to the crime of murder and those convicted of it, but they also exceeded and escaped control of the state and took on new meaning and power.

NOTES

1. Before the Murder Act, those of the nobility condemned for a capital crime were executed by beheading which was considered in line with their status. Execution under the Murder Act was a much more egalitarian affair: regardless of status, all convicted under the act were sentenced to 'hang by the neck until dead'.
2. See, Gatrell, V.A.C. (1994), *The Hanging Tree: Execution and the English People 1770–1868* (Oxford: Oxford University Press), p. 201.
3. See, Foucault, M. (1975), *Discipline and Punish: The Birth of the Prison* (New York: Vintage Books); Ignatieff, M. (1978), *A Just Measure of Pain: The Penitentiary in the Industrial Revolution, 1750–1850* (New York: Pantheon Books).
4. For more on these developments, see Ward, R. (2014), *Print Culture, Crime and Justice in Eighteenth-Century London* (London: Bloomsbury), p. 162.
5. See, Ward, R. (2014), *Print Culture, Crime and Justice in Eighteenth-Century London* (London: Bloomsbury), p. 167.
6. Ibid., p. 171.
7. Ibid., p. 202.
8. See, Ward, R. (2012), 'Print Culture, Moral Panic, and the Administration of the Law: The London Crime Wave of 1744', *Crime, History & Societies*, Vol. 16, 5–24.

9. On the longer history of hanging in chains in Britain, see Tarlow, S. *The Golden and Ghoulish Age of the Gibbet* (Palgrave 2017). Although some criminal corpses were given over to the medical men by the authorities before the advent of the Murder Act, dissection was not intended as a punishment and part of achieving the aims of the criminal justice system. The families of those whose bodies were given to the medical men under royal or other grants probably saw this as an additional punishment nonetheless.

10. See, Ward, R. (2015), 'Introduction', in Ward, R. (ed.), *A Global History of Execution and the Criminal Corpse* (Palgrave Macmillan), p. 24.

11. See, Ward, R. (2014), *Print Culture, Crime and Justice in Eighteenth-Century London* (London: Bloomsbury), p. 158.

12. See, Linebaugh, P. (1975), 'The Tyburn Riot Against the Surgeons', in Hay, D., Linebaugh, P., Rule, J.G., Thompson, E.P., and Winslow, C. (eds.), *Albion's Fatal Tree* (New York: Pantheon Books), pp. 65–118.

13. See, Gatrell, V.A.C. (1994), *The Hanging Tree: Execution and the English People 1770–1868* (Oxford: Oxford University Press).

14. See, Hurren, E.T. (2016), *Dissecting the Criminal Corpse: Staging Post-execution Punishment in Early Modern England* (London: Palgrave Macmillan); King, P. (2017), *Punishing the Criminal Corpse 1700–1840: Aggravated Forms of the Death Penalty in England* investigates more direct possible applications of the principle of *lex talionis*: some commentators in the 1740s and 1750s advocated that retaliatory harm be inflicted on the condemned before execution as retribution for the violent crimes they committed, and that the form of that punishment follow what the condemned had done to their victim. One writer 'optimistically believed that forcing the condemned through the "same process of pain and horror" as the victim would have a preventative role, because "the deliberating villain, designing the murderous blow, would from a sudden recollection that he might afterwards feel the same painful stroke … stay his hand in the work of horror" (King, *Punishing the Criminal Corpse* quoting *Old England or The National Gazette*, 1 February 1752).

15. Davies and Matteoni noted that breaking on the wheel was being used in the German states of Prussia and Hesse-Kassel into the nineteenth century. On this, and for a description of this punishment, see Davies, O. and Matteoni, F. (2017), *Executing Magic: Criminal Bodies and the Gallows in Popular Medicine and Magic in the Modern Era* (Palgrave), p. 2.

16. See, King, P. (2017), *Punishing the Criminal Corpse 1700–1840: Aggravated Forms of the Death Penalty in England*, chapter 2.

17. Murder Act.

18. See, Banner, S. (2002), *The Death Penalty* (Cambridge, MA: Harvard University Press), p. 81.

19. See, Ward, R. (2014), *Print Culture, Crime and Justice in Eighteenth-Century London* (London: Bloomsbury), p. 183.
20. See, Ward, R. (2015), 'Introduction', in Ward, R. (ed.), *A Global History of Execution and the Criminal Corpse* (Palgrave Macmillan), p. 10.
21. They also differed in important ways, as is discussed in Chapters 5 and 6.
22. Bernard Shaw, G. (2012 [1903]), *Maxims for Revolutionists* (CreateSpace).
23. On pardoning rates for capital crimes in this period, see King, P. (2000), *Crime, Justice and Discretion in England 1740–1820* (Oxford: Oxford University Press). On pardoning of those convicted under the Murder Act, see King, P. (2017), *Punishing the Criminal Corpse 1700–1840: Aggravated Forms of the Death Penalty in England*, chapter 3.
24. Old Bailey Online, Ordinary's Account, Thomas Wilford, Killing, murder, 25 June 1752, https://www.oldbaileyonline.org/browse.jsp?id=t17520625-31-off133&div=t17520625-31#highlight.
25. See, Bennett R. (2018), *Capital Punishment and the Criminal Corpse in Scotland, 1740–1834* (London: Palgrave Macmillan).
26. See, King, P. and Ward, R. (2015), 'Rethinking the Bloody Code in Eighteenth-Century Britain: Capital Punishment at the Centre and on the Periphery', *Past and Present*, Vol. 228, 159–205, 183.
27. See, Hurren, E.T. (2016), *Dissecting the Criminal Corpse: Staging Post-execution Punishment in Early Modern England* (London: Palgrave Macmillan), p. 21.
28. Pardons given to those convicted of murder almost always consisted of a reprieve from execution. As Peter King has noted, it was 'extremely rare for the post-execution part of the sentence to be formally removed unless the offender had also been pardoned from the death sentence itself.' (King, P. (2017), *Punishing the Criminal Corpse 1700–1840: Aggravated Forms of the Death Penalty in England*, chapter 3.)
29. See, Gatrell, V.A.C. (1994), *The Hanging Tree: Execution and the English People 1770–1868* (Oxford: Oxford University Press).
30. See, Poole, S. (2016), '"For the Benefit of Example": Crime-Scene Executions in England, 1720–1830', in Ward, R. (ed.), *A Global History of Execution and the Criminal Corpse* (London: Palgrave Macmillan), p. 81.
31. Ibid., p. 78.
32. Ibid., p. 84.
33. See, Laqueur, T.W., (1989), 'Crowds, Carnival and the State in English Executions, 1604–1868' from Beier, A.L., Cannadine, David and Rosenheim, James M. (ed.), *The First Modern Society: Essays in English History in Honour of Lawrence Stone* (Cambridge: Cambridge University Press), pp. 305–355.
34. See, Devereaux, S. (2009), 'Recasting the Theatre of Execution: The Abolition of the Tyburn Ritual', *Past and Present*, 172, 202.

35. See, King, P. and Ward, R. (2015), 'Rethinking the Bloody Code in Eighteenth-Century Britain: Capital Punishment at the Centre and on the Periphery', *Past and Present*, Vol. 228, 159–205.
36. Ibid., 159–205, 194.
37. Ibid., 159–205, 170.
38. See, MacDonald, M. (1986), 'The Secularization of Suicide in England, 1660–1800', *Past & Present*, Vol. 111, 50–100, 53.
39. See, Kästner, A. and Luef, E. (2015), 'The Ill-Treated Body: Punishing and Utilizing the Early Modern Suicide Corpse', in Ward, R. (ed.), *A Global History of Execution and the Criminal Corpse* (London: Palgrave Macmillan), pp. 147–169, p. 147.
40. See, MacDonald, M. (1986), 'The Secularization of Suicide in England, 1660–1800', *Past & Present*, Vol. 111, 50–100, 53–54.
41. See, Cherryson, A. Crossland, Z., and Tarlow, S. (2012), *A Fine and Private Place* (Leicester: Leicester Archaeological Monographs), p. 119. An alternative folkloric explanation is the crossroads, by their form, are self-consecrated, and thus burial at a crossroads could be construed as an act of mercy.
42. See, MacDonald, M. (1986), 'The Secularization of Suicide in England, 1660–1800', *Past & Present*, Vol. 111, 50–100.
43. See, Kästner, A. and Luef, E. (2015), 'The Ill-Treated Body: Punishing and Utilizing the Early Modern Suicide Corpse', in Ward, R. (ed.), *A Global History of Execution and the Criminal Corpse* (London: Palgrave Macmillan), pp. 147–169, p. 151.
44. See, MacDonald, M. (1986), 'The Secularization of Suicide in England, 1660–1800', *Past & Present*, Vol. 111, 50–100, 98.
45. Ibid., 50–100, 78.
46. For example, in a survey of suicide in the six northern counties over fifty years based on *The Cumberland Pacquet*, R.A. Houston found eighteen reports of *felones de se*, of which three mention unusual burials (1790-1791). See, Houston, R.A. (2010), *Punishing the Dead? Suicide, Lordship, and Community in Britain, 1500–1830* (Oxford: Oxford University Press), p. 202.
47. Old Bailey Proceedings Online (www.oldbaileyonline.org, version 6.0, 14 September 2012), Ordinary of Newgate's Account, September 1760 (OA17600915). Ordinary's Account, 15 September 1760. Francis Stirn, Murder, Suicide then Dissected.
48. WEP 13 September 1760.
49. Assize Calendar E389/246/77d.
50. Gloucester Journal March 1777.
51. See, Cobb, R. *Death in Paris: The records of the Basse-Geôle de la Seine, October 1795–September 1801* (Oxford 1978), p. 101.

52. See, Kästner, A. and Luef, E. (2015) 'The Ill-Treated Body: Punishing and Utilizing the Early Modern Suicide Corpse', in Ward, R. (ed.), *A Global History of Execution and the Criminal Corpse* (London: Palgrave Macmillan), pp. 147–169, p. 151.
53. E389/246/77d.
54. LC 10 September 1791.
55. Sheriff's cravings: TNA T90/167 Birch's body 'to be buried in the cross roads near Tewkesbury as directed by the judge—he having committed an act of suicide after being convicted of murder'.
56. *Ipswich Journal*, 31 March 1804.
57. See, Healy, R. (2006), 'Suicide in Early Modern and Modern Europe', *The Historical Journal*, Vol. 49, Issue 3, 903–919. Róisín Healy discusses other European contexts in which the modern states responded to the threat suicide posed to the state monopoly on violence by seeking to prevent suicides after decriminalisation.
58. *Times*, 7 February 1785 (see, Ward, R. (2015), 'The Criminal Corpse, Anatomists and the Criminal Law: Parliamentary Attempts to Extend the Dissection of Offenders in Late Eighteenth-Century England', *Journal of British Studies*, Vol. 54, 63–87, 67).
59. See, Ward, R. (2015), 'The Criminal Corpse, Anatomists and the Criminal Law: Parliamentary Attempts to Extend the Dissection of Offenders in Late Eighteenth-Century England', *Journal of British Studies*, Vol. 54, 63–87, 78 (quoting Loughborough in: HOC Papers. PR; 5 July 1786, p. 160).
60. See, Ward, R. (2014), *Print Culture, Crime and Justice in Eighteenth-Century London* (London: Bloomsbury), p. 187.
61. Ibid., pp. 186–187.
62. See, Anderson, C. (2015) 'Execution and Its Aftermath in the Nineteenth-Century British Empire', in Ward, R. (ed.), *A Global History of Execution and the Criminal Corpse* (Palgrave Macmillan), p. 171.
63. See, Ward, R. (2012), 'Print Culture, Moral Panic, and the Administration of the Law: The London Crime Wave of 1744', *Crime, History & Societies*, Vol. 16, 5–24.
64. See, Ward, R. (2015), 'The Criminal Corpse, Anatomists and the Criminal Law: Parliamentary Attempts to Extend the Dissection of Offenders in Late Eighteenth-Century England', *Journal of British Studies*, Vol. 54, 63–87, 63.
65. Ibid., 63–87, 66.
66. Ibid., 63–87, 78 (quoting Loughborough in: HOC Papers. PR; 5 July 1786, 160).
67. Ibid., 63–87, 79 (quoting HOC Papers. PR; 11 March 1796, 287).
68. Ibid., 63–87.

69. Ibid., 63–87, 80 (quoting Chief Justice of Chester, James Adair, HOC Papers, PR, 11 March 1796, 289).
70. See, King, P. (2017), *Punishing the Criminal Corpse 1700–1840: Aggravated Forms of the Death Penalty in England* (London: Palgrave Macmillan).
71. Ibid., chapter 4.
72. See, Davies, O. and Matteoni, F. (2017), *Executing Magic in the Modern Era: Criminal Bodies and the Gallows in Popular Medicine* (Palgrave Macmillan), p. 3.
73. See, Richardson, R. (2000), *Death, Dissection and the Destitute* (Chicago: University of Chicago Press, 2nd Edition) for the Anatomy Act and its impacts.

CHAPTER 5

Anatomisation and Dissection

Once opportunities for clemency or escape—a pardon, commuted sentence, or other reprieve—had passed, capital conviction meant one thing: execution. However, death at the end of a hangman's rope, while often taken as a clear conclusion in studies of crime and punishment, was not the end of the judicial process, nor the end of the criminal's narrative journey or their capacity to play a powerful and meaningful role in the social, scientific and cultural life of the nation.[1] As we have seen, power inhered in the criminal body far beyond the spectacle and moment of execution, whether in terms of persistent vitality, such as instances of asynchronous legal and medical death,[2] or as an object giving rise to fear, fascination, disgust and desire.[3] The criminal corpse was the locus of new spectacles of state power, post-mortem punishments shaped by retributive justice, and modern scientific experimentation. Further, the living prisoner and the criminal corpse did not exist in a neat dichotomy, one becoming the other thanks to a short drop and sudden stop. As medical men worked with the bodies they received via the Murder Act, it became clear that the lines between life and death—and consequently, between live prisoner and criminal corpse—were far less clear than might be expected. This chapter picks up the story of the criminal corpses produced through convictions and executions under the Murder Act, specifically of those sentenced to the post-mortem punishment of anatomisation and dissection, and follows the fates of these bodies as

© The Author(s) 2018   115
S. Tarlow and E. Battell Lowman, *Harnessing the Power of the Criminal Corpse*, Palgrave Historical Studies in the Criminal Corpse and its Afterlife, https://doi.org/10.1007/978-3-319-77908-9_5

they moved through the nexus of medical knowledge, dismemberment, public spectacle, death and decay.

The criminal corpse is our focus here, but we must also ask: who were the people working with these bodies, and why did they choose to do so? The answer begins by noting the close tie between the Murder Act and the changing status of medical dissection. By stipulating that the body of a murderer 'shall be dissected and anatomized by the said *Surgeons*, or such person as they shall appoint for that purpose',[4] the Murder Act created an official role for medical professionals in the British criminal justice system. Until 1745, the Company of Barber-Surgeons essentially held a monopoly on accrediting surgeons, but this monopoly was hardly a barrier to calling oneself a surgeon. Many men with a variety of accreditations, or in fact no accreditation at all, commonly claimed to be surgeons. This began to change in 1745 when the Company of Surgeons was formed in London following a long-anticipated split from the Company of Barber-Surgeons. In 1752, 'surgeon' became a standardised qualification, and the Company of Surgeons gained the largely exclusive power of accreditation. There is a further divide in the application of the term between those who sought and gained this accreditation for the purpose of general practice—called 'apothecary-surgeons'—and those who conducted surgery specifically within the penal system—'penal surgeons'. This is an important distinction. An apothecary-surgeon required accreditation, and many apothecary-surgeons also served as penal surgeons, but not all penal surgeons were so accredited and they did not all work as apothecary-surgeons. This was especially true outside of London, where surgeons were scarcer, and the presence of the Company of Surgeons somewhat more distant. These penal surgeons were often men who held some other medical experience or credential, and were locally respected for their knowledge and skills.

Surgeons were not the sole agents of the state involved in conducting anatomisation and dissection. Rather, a broader medical community was involved in fulfilling this role under the Murder Act, including physicians, students and the paid staff of the Company (such as porters and beadles). Given the imprecision of the term 'surgeon' in this period, and the numerous other actors involved in carrying out this post-mortem punishment,[5] it is perhaps most accurate to speak of 'medical men': a largely (if not exclusively) homosocial group of adherents to the burgeoning medical-scientific complex, ranging from experts in human anatomy, to clerks and craftspeople who can be understood as stakeholders with practical investment in the judicial process. It was these medical men—of all stripes—who were critical to carrying out the sentence of anatomisation and dissection and staging the public and professional spectacles that followed death on the gallows.

Historical literature has paid scant attention to post-execution rites, though the spectacle and process of execution has been of considerable interest to historians of Britain's long eighteenth century.[6] However, more by omission than deliberate neglect, this created the mistaken impression that penal surgeons handled only 'dead bodies from the gallows and that capital penalties from a medical standpoint were straightforward once a criminal stopped jerking on the hangman's rope'.[7] What happened after the sentence of legal death was accomplished on the gallows is much more complex, blurring the lines between life and death, and giving rise to a whole suite of post-execution rites, processes and spectacles. This chapter takes up the journey of the criminal corpse from the foot of the scaffold and into the spaces of the first of the post-mortem punishments mandated by the Murder Act: anatomisation and dissection.

DUTY, DEATH, AND DISCRETION

Under the Murder Act, surgeons (and within Middlesex and London, the Company of Surgeons) were charged with the duty of anatomising and dissecting the corpses of executed murderers sentenced to suffer these 'marks of infamy'.[8] Anatomisation in this context refers to an established process of opening the corpse and checking vital organs—the heart and lungs up to 1812, and the heart, lungs, and brain thereafter—in order to establish death with certainty.[9] Dissection in this context is best understood as the further infliction of post-mortem harm on the body for medical training and research purposes. Anatomisation including displaying its results to the execution crowd and dissection required medical men to take a leading role in both the practice and the public display of this post-mortem punishment.

Though medical men had long been peripherally related to the criminal justice system—tending sick prisoners who could afford treatment while in gaol, and obtaining the pre-Murder Act bodies allotted them from the gallows for dissection in the service of medical training—the Murder Act for the first time made medical professionals formal actors in the British criminal justice system.[10] As surgeons were made responsible for executing a key stage in the punishment of those deemed society's worst offenders, the connection between medicine and criminal justice became fixed in the public imagination. Over the course of the eighteenth century, extensive coverage in the popular press, street ballads, and other entertainments built and hardened the association of executions with medical training and professionalisation.[11] The presence of the medical men at executions and public dissections, performing both ceremonial and practical public functions, allowed average folk to confirm these associations with their own eyes. This helped to generate powerful and enduring impressions of medical men as both agents of the state, and of death, at times to their dismay.[12]

It is easy to draw a simple association between medical men and execution and dissection under the Murder Act, but the degree to which surgeons and others could exercise agency within and even against the juridical regime created by the Act is not obvious. The Act is remarkably clear in stipulating some matters related to sentencing and punishing convicted murderers, including the conditions of confinement of the condemned before execution, the timing of sentencing and execution, and the choice between two mandated post-mortem punishments. The Act is far less clear when it comes to key elements of the sentence of dissection and anatomisation. It did not stipulate where and when the procedures should take place, who should be present, or how long the punishment should last, instructing only that the body be taken to the appointed surgeon, and that 'in no case whatsoever the body of any murderer shall be suffered to be buried; unless after such body shall have been dissected and anatomized'.[13] This created both the space and necessity for those involved in executing the sentence of post-mortem punishment to develop protocols of their own through practice and example. The medical men had to determine in carrying out this post-mortem punishment: how publicly visible their work would be, what types of anatomical techniques they would employ, how much of a body would be left afterward, what parts might be kept and preserved, and how, when, and where the remains would be disposed of.[14] Sometimes, as we shall see, this included life-and-death decisions.

'I'm Not Dead Yet!' Medical Men and the Uncertainty of Death

The punishment for capital crimes in eighteenth and nineteenth century Britain was to be hanged by the neck until dead. But death on the gallows was no easy thing. Those of us more accustomed to depictions of hanging in television and films than historical and actual hangings likely have a distorted idea of how this form of execution looks, kills, or smells.

Under the procedures in place during the time of the Murder Act, death on the gallows was never totally certain—legally or otherwise.[15] We are likely more familiar with the clinical precision of the ideal 'long drop' in which the upper cervical vertebrae are quickly fractured or dislocated when the body's acceleration as it falls is stopped short by the noose, the sudden jerk and resulting trauma to the neck causing immediate unconsciousness and rapid death.[16] But this innovation, and the speedy death it promised, was not introduced until well after the period of the Murder Act.[17] Instead, the 'short drop' was the method used to hang those sentenced to die in Britain in the eighteenth and early nineteenth centuries.[18]

In Britain, the condemned were taken to the gallows with their arms tied; a cap was placed over their face and the noose placed around their neck. Then, the cart, ladder, or trapdoor on which they stood was removed, leaving them to dangle by their neck at the end of the taught rope. Usually the individual began to die of strangulation, their skin stretching under the weight of their body, their neck dislocating.[19] Should the neck not break, the restriction of blood flow created incredible pressure inside the head, resulting in protruding eyes, the face turning vivid purple, then black, and the brain turning into a 'bloody mush'.[20] The pressure and trauma caused the body to evacuate. Faeces and urine were joined by sex-specific discharges—in men, the release of seminal fluid and in women, spontaneous menstruation as the uterus prolapsed.[21] Death by the short drop was excruciatingly painful and unavoidably messy.

The effectiveness of this method of execution depended a great deal on the hangman's individual skill but also on other factors, some difficult to overcome. Ideally, bodies were left to hang for an hour to ensure death. In newspaper reports on hangings during this period, this was described as leaving the body to hang 'for the usual time'. However, in summer months, the heat made it unpleasant to leave a body hanging for the full hour. Conversely, cold temperatures in winter could send a body into hypothermic shock, slowing life signs so that it was difficult to determine if death had occurred, necessitating longer waits. Further, perspiration from fear or heat could make the noose slip and slide and affixing the noose in the most effective way could be equally difficult if prisoners struggled. Finally, the physical attributes of the condemned could make hanging more difficult or prevent death on the gallows. The 'bull necked' posed a particular problem as strong muscles could protect the arteries in the neck from being constricted by the rope, allowing (some) blood flow to continue.[22] Though implicit, it was also clear that 'the appointed executioner was duty-bound to ensure that the condemned died on the rope'.[23] However, the capital code did not allow the use of bladed weapons to finish off the condemned so the hangman, or family and friends of the dying, could only resort to handling the body more roughly—in particular, tugging on the legs—to ensure or speed up death by strangulation or broken neck if this was thought necessary.[24]

With so many factors influencing the effectiveness of short drop hanging, it is no surprise that not everyone brought down from the gallows was dead. In cases sometimes referred to as the 'half-hanged', individuals revived after hanging.[25] Brenda Cook has identified 13 instances of individuals surviving execution by hanging and reviving afterward in Britain between 1587 and 1785, and of these 2 were immediately re-hanged, and 5 died of the injuries sustained from their mandated punishment.[26] Though remarkable and very well reported in the press, revival after execution was atypical. Much more common was finding that a body brought down from the gallows, though incapable of revival, was evidently not yet completely dead.

The issue of death, or rather the uncertainty of determining death, was well discussed in medical circles in the two centuries before the advent of the Murder Act. The combination of a less than one hundred percent effective method of execution with complicating factors such as weather or particularly robust physiques, or variations in the amount of time a body was left to hang, meant that some of those sentenced to death for murder actually died elsewhere in the presence of, or at the hands of, the surgeons. Accounts of anatomists beginning dissections on bodies thought dead that subsequently—and sometimes, spectacularly— revived were widely circulated, including being retold in J.B. Winslow's instructions on responsibilities, timings, and techniques for medical men in his important 1746 volume, *The Uncertainty of the Signs of Death, and the Dangers of Precipitate Interments and Dissections, Demonstrated.*[27] In one well-known case from the sixteenth century, Winslow mentions that the anatomist began cutting into the 'corpse' provided to him, only to discover the dead person was in fact still alive—but not for long, as the anatomist's initial cuts completed the job. Consequently, the anatomist was chased out of town by enraged members of the public shouting

'Murderer!' The pursuit of medical knowledge was not without risk, a fact of which the medical men of the eighteenth and nineteenth centuries were only too aware, and a point to which we return later in this chapter.

In the years immediately preceding the Murder Act, Winslow noted the difficulty of determining death in cases of hanging as 'we are often deceived with respect to the Signs of Death' and many of the ways death might be identified—including the colour of the face, flexibility of the limbs, temperature of the body, and the 'abolition of the external senses'—are 'very dubious and fallacious Signs of a Certain Death'.[28] Winslow concluded that the only truly infallible way to determine death was to wait for decomposition to begin. This was contrary to the needs of anatomical work, but the moral and ethical orthodoxy of the day held that it was better to wait than to accidentally cause death with the surgeon's blade. The celebrated anatomist Jean Riolan (the Younger, 1577–1657) specifically addressed the issue of uncertainty of death and state-sanctioned dissection:

> [S]peaking of the Bodies of hanged Persons, by public Authority destined to Dissection... That so long as the Body is warm, and the Person but lately executed, we are not to dissect him; since, if there is still any Prospect of recalling him to Life, we are equally bound by the Principles of Humanity and Charity to do all we can for that Purpose, in order to procure him, if possible, a favourable Opportunity of Repentance.[29]

Riolan prioritised the preservation of life and the avoidance of foreclosing on any opportunity for a person to 'die properly'—that is, to be given the opportunity for repentance and thus salvation—over and above the potential anatomical benefits of beginning a dissection quickly in order to make use of the body while it was as fresh, and therefore as useful, as possible. In this way, Riolan effectively advocated yielding to the importance of the time of the dead, instead of anatomical time. Winslow noted that Terilli, the celebrated early seventeenth-century physician of Venice, was even stronger in his call to delay dissection until true death could be confirmed, and the imperative for the medical men to yield to the time of the dead, because:

> [The Body] is sometimes so depriv'd of every vital Function, and the Principle of Life reduc'd so low, that it cannot be distinguished from Death, the Laws both of natural Compassion and reveal'd Religion oblige

us to wait a sufficient Time for Life's manifesting itself by the usual Signs, provided it should not be as yet totally extinguished; and if we should act a contrary Part, we may possibly become Murderers, by confining to the gloomy Regions of the Dead, those who are actually alive.[30]

This concern about the anatomist-as-murderer in cases where bodies reached medical men before life had completely left the body led Winslow to argue that the best practice was to leave the supposedly dead individual supine with a pillow under the head and covered by a blanket, and to wait two or three days. By this time, either a return to life or an incontrovertible death would have taken place, and one imagines that after three days death could be easily confirmed by smell alone.[31]

In the case of those executed under the Murder Act, the 'time of the dead' was not held sacrosanct in the way Winslow and Riolan might have preferred. Elizabeth Hurren writes about the case of John Holloway, sentenced to death and dissection in 1831 for the 'horrible murder, almost unparalleled in atrocity'[32] of his wife, Celia Holloway. Being strong of neck, Holloway was considered a 'dangerous' body because even after an hour on the scaffold, his neck wasn't broken, meaning there was a risk that he might revive. Hurren writes that 'The body now had to be made safe by the surgeon' by severing the carotid artery (in the neck) to speed up the dying process.[33] Coming towards the end of the life of the Murder Act, we now know that this instance of a surgeon assuring or causing the medical death of the condemned was no isolated incident. Hurren has found a startling number of cases in which criminal corpses received by the medical men were not in a state of absolute death. The records of William Clift, who worked at Surgeon's Hall in London show that between 1812 and 1830, of 35 well-documented cases, there were 10 in which the condemned was not yet medically dead, that is 'the heart was still beating after the body was received'.[34] Did Clift and others in his place follow the strong calls by Riolan, Terilli, and Winslow to wait for absolute death before proceeding? In a word: no.

It was not just common knowledge but also a generally unchallenged practice that medical men might end the lives of condemned criminals, despite the prohibition on completing the work of the noose with a blade. As a newspaper correspondent wrote in 1769: 'the business of Surgeon's Hall is not to revive and frustrate but to complete the Execution of the Sentence in Cases of Murder'.[35] In the case that the body of a hanged murderer delivered to the medical men under the Murder Act showed any sign of life, the first duty of the surgeon was to 'use the lancet to commit a merciful act'[36]—that is, to supplement the hangman's rope with the penal surgeon's lancet to complete the transformation from condemned to corpse. In large part this was a result of the challenges in the use of short-drop hanging as the exclusive method of execution during this period, but it was also because of the great difficulty in distinguishing between the two physiological types of death identified in the eighteenth and nineteenth centuries: 'the name of death', in which a body was unresponsive to stimuli, and 'absolute death', signalled by a complete physical shutdown.[37] We will return to the issue of determining death, and the work the medical men conducted with bodies in the state between the name of death and absolute death in the final section of this chapter; what we want to underscore here is the role of the medical men in relation to the criminal justice system. The Murder Act refers to the surgeon as being responsible for anatomising and dissecting the bodies of those sentenced under the Act. Unofficially, but indisputably, the surgeon was also responsible for causing or hastening death. In some cases, they were co-executioners.

Between Science, Spectacle and the State

Under the terms of the Act, those murderers not sentenced to hang in chains were sentenced to anatomisation and dissection. The letter of the law appears straightforward here, particularly as the two terms were (and in some cases, still are) used interchangeably. The bottom line was that the surgeons would cut the murderer's corpse, and that burial was not permitted until this had taken place. In practice, however, carrying out this sentence was anything but clear-cut once the medical men were in possession of a body. For the surgeons, their actions were dictated not only by the law, but also by the execution crowd. Both of these external pressures were further affected by the personal and professional capabilities and priorities of the medical men themselves. Execution crowds in eighteenth- and nineteenth-century Britain were regularly thousands-strong. Drawn by the spectacle of punishment, by the frisson of excitement, the carnivalesque crowds were at once attracted to and repulsed by the visceral display, responding to life, death, authority, each other and the criminals themselves.[38] Thomas Laqueur has argued that the crowd 'was the central actor in English executions',[39] while Peter Linebaugh has established that the crowd was able to exercise a significant amount of power through strong, collective reactions to anatomists at work.[40] However, studies to date have usually left the crowd (along with the corpse) at the gallows. By intent or omission this neglects the significant power of the execution crowd in relation to punishment that did *not* end at the hanging tree.

One of our key findings in tracing the journey of the criminal corpse is that the post-execution crowd was a key actor in determining the location and extent of public post-mortem punishment of those convicted under the Murder Act.[41] Post-execution, the crowd also expected to see the post-mortem punishment *and* to participate. The reasons for the crowd's interest in witnessing and participating in the punishment of the corpse are similar to the reasons for attending the execution. However, we argue that four key elements fuelled the crowd's interest in seeing the body of a murderer opened and exposed: the urge to see that justice had been done and the evildoer was well and truly dead (with no risk of resurrection); 'natural curiosity' about the dangerous dead, as murderers and other criminals were often at the centre of news, gossip, and local folk tales; curiosity about the shaved, nude, fleshy body which was otherwise rarely seen in public, including a particular interest in the sexual organs which, in the case of hanged men, may have been in a state resembling excitement; and the prestige of proximity and witnessing a well-known event that would become part of history, granting the participant the right to declare 'I was there!' The post-execution crowd clamoured for access, and to deny them was dangerous and difficult, if not impossible.

Though not written into the Murder Act or specifically mandated by the criminal justice system, making the post-mortem punishment of anatomisation and dissection a public event did serve State interests. As discussed in the previous chapter, the Act was intended to deter potential murders by inspiring horror at the prospect of the dismembering and decay of their body and the denial of respectable burial and its associated rites. The shame and humiliation of public dissection supported this end, as evidenced by tales of prisoners described as stoic during the pronouncement of execution, but who lost their nerve at the prospect or sentencing of post-mortem dissection.[42] Further, in the eighteenth and early nineteenth centuries, justice had to be done locally, which is to say that justice had to be meted out in front of local audiences for justice to be *seen to be done*. Prior to the advent of mass media, news tended to circulate regionally, and it was difficult to separate fact from fiction as stories travelled across distances and in many versions, and were interpreted in a range of class, cultural and personal contexts.[43] Seeing was—quite literally—believing when it came to the delivery of justice. While it is true and not inconsequential that the state benefitted from public post-mortem punishment in the way that it increased the terror and infamy of the punishment for murder, it was the crowd that drove the public imperative. For example, the courtyard of the Shire Hall in Derby was altered in 1752 to permit the crowd a better view of executions and the transportation of the corpse from the gallows into the room used for dissection. Railings and gates were added (through which the crowd entered to walk around the displayed corpse) in the hope that by better permitting controlled visual and physical access that the crowd would be satisfied and less likely to riot.[44] Clearly, the crowd did not lose interest in the criminal or the punishment spectacle once it could be called a corpse. Rather, public interest remained strong and as such, spectators had to be accommodated in two senses: they had to be allowed *physical access* to view the bodies opened by the surgeons, and they had to be allowed *conceptual access* in that the post-mortem punishment had to meet their expectations of such a spectacle so that they would be satisfied, and disperse.

To allow the crowd physical access to see the anatomised criminal corpse, the body had to be brought to a place where it could be displayed and people could see it, usually by moving past the corpse in long queues. For this reason, a variety of public and semi-public spaces close to the site of execution were used as dissection and anatomisation venues.[45] Hurren has identified four broad types of spaces used for this purpose.[46] In the north of England to the west of the Pennines, it was common to use small public dispensaries; in the Midlands, the local Shire Hall was a preferred site; in London, before criminal corpses were taken to central locations for private anatomical work, public anatomisation— particularly in the cases of very violent murderers—was sometimes conducted at the site of the crime to increase the symbolic impact of the punishment[47]; finally, in the West Country, post-mortem punishment of murderers was carried out in prison rooms, the domestic premises of the surgeon, or in a medical dispensary.[48] Indeed, Hurren's central finding is that 'post-mortem "harm" was always located in public spaces in which it would gain greater acceptance by a wide cross-section of the community'.[49] The local and accessible nature of post-mortem punishment under the Murder Act was necessary to effectively convey the importance of punishing murderers in particularly degrading or torturous ways, a component of expanding social control as discussed in the previous chapter. Making criminal anatomisations local and highly visible during the same period that the actual numbers of capital convictions for murder were declining significantly increased the impact and the scope of influence of the punishment for murder. During this period the State increasingly worked to limit interaction between the crowd and medical men at the site of execution. The reverse was true in the places and spaces where the post-mortem punishment was conducted, where the public was encouraged to crowd around, cheer and jeer, and otherwise turn the site into a macabre carnival.[50] State actors knew quite well that, for post-mortem punishments to have an impact on the public, they had to be conducted in a place accessible to large numbers.

In many respects, the requirement for the crowd to have physical access to the post-mortem spectacle dictated the parameters for the spaces in which the medical men could execute their duty under the Act—large enough and central enough to house the excited onlookers, while also providing a clear, central space for the medical men to stage or work on the corpse.[51] However, the conceptual access required by the crowd also influenced the anatomical procedures the medical men chose to perform during their work. The anatomical procedures that supported the teaching and research needs of the medical men were impossible in the context of demonstration for the post-execution crowd. A lack of sanitation and contamination of the body were common issues, as was the lack of light, quiet, appropriate storage facilities, and specialised tools that might be difficult or impractical to transport. Further, the uses to which the medical men wanted to put these bodies did not necessarily align with the expectations of the crowd. They wanted to see a body that still looked like a body and this was impossible should the medical men proceed in accordance with their teaching and research interests, often meaning detailed examination of particular organs and dissection 'to the extremities'. This type of dissection resulted in only about a third of the body remaining intact, as organs were removed and further dissected, flesh and cartilage removed from bone, circulatory systems carefully disassembled, reassembled and mapped, and various other disintegrations.[52]

Surgical anatomists already had a poor public reputation, stemming from the grisliness of their long-standing use of corpses granted them from the gallows, and their reputation for illegally buying bodies stolen from graves. Upsetting the crowd's expectations could mean intense reprisals, including being chased from town by an angry mob, and more usually vandalism or damage to the places used to display or conduct anatomisation and dissection.[53] The medical men, therefore, had to develop practices that allowed the negotiation between their duty in the criminal justice system, the expectations of the crowd, and the exigencies of their own interests in these corpses.

The medical men engineered a balance by disambiguating their actions into two distinct processes—anatomisation and dissection—and by working specifically to satisfy the expectations of the crowd so that they could conclude the public and criminal justice aspect of their work as quickly as possible and repair to their anatomy rooms with the corpse while it was still in a useful state, before significant decay had begun. While the full process of anatomisation and dissection might differ, as did the meaning and use of these terms between people and regions, penal surgeons frequently used them interchangeably to refer to the cutting of a body to reveal or allow the study of internal structures and the workings of the body, and to confirm death. This took advantage of ambiguity in the wording of the Murder Act, which referred clearly to anatomisation and dissection, but made no distinction between the two, allowing medical men to develop their own clarifications to suit their multiple needs.[54]

Given the primacy placed on ensuring a criminal had become a corpse before truly committing to dissection for both punishment and educational purposes, the confirmation of death through anatomisation became the first duty. This procedure involved checking to ensure the body was now lifeless and confirming this for the crowd by putting it on display for the public to see.[55] In order to achieve the desired effect, including displaying a body that was still recognisable and showed harm had been done to the corpse, the surgeons typically made cuts deep enough to check that the heart and lungs (and later, brain) had lost function, but not necessarily involving the mass opening or removal of large pieces of flesh, such as the breast plate and rib cage or whole limbs. Anatomisation usually involved making two intersecting cuts—one from about neck to groin, and a perpendicular cut across the chest or abdomen—to permit manual and some visual access to the main organs. As one example, the anatomisation of William Corder, executed in 1828 for the murder of Maria Martin, was described in detail in newspaper reports[56]:

> Mr. Creed, the county surgeon, assisted by Mr. Smith and Mr. Dalton, made a longitudinal incision along the chest as far as the abdominal parts, and deprived it of the skin so as to exhibit the muscles of the chest.[57]

These cuts allowed the surgeons to slice into the main muscles, peel back the skin, and expose to view the organs (possibly removing some). The body, stripped of all or most clothing—in the case of men doubly exposed through shaving[58]—and presented with long cuts exposing muscles and organs for display fulfilled the public's idea of what medical dissection should look like while simultaneously presenting an identifiable criminal who had clearly received the mandated punishment, all while preserving as much of the body as possible for later anatomical use (Fig. 5.1).

Dissection allowed the surgeons much greater latitude to pursue their own priorities with respect to the criminal corpse. First, this usually involved moving the body to a venue better suited to medical work and

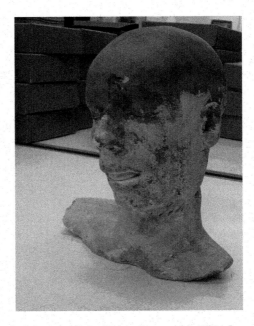

Fig. 5.1 Cast of a hanged criminal's head, owned by Winchester Museums (Photo: S. Tarlow)

then conducting as many as seven different anatomical procedures, until 'the murderer was despoiled as a human being'.[59] Dissection, then, was the element of the punishment mandated in the Murder Act that allowed medical men to make use of criminal corpses in ways much more closely aligned with their own priorities and needs, and without the same imperative to play to an audience and maintain public perceptions of what an executed and punished murderer should look like. To illustrate this, we return to the post-mortem punishment of William Corder: once anatomisation and the public access and viewing involved was complete, the corpse was moved from the Shire Hall to the County Hospital. Here, the same surgeons who had performed the anatomisation dissected before an audience of 'medical gentlemen':

> Mr. Creed, jun. assisted by Mr. C. Smith, and Mr. Dalton, commenced the operations; they first minutely dissected the muscles of the chest, and having elevated the sternum, and examined the lungs, they took out the intestines, all of which appeared in a most healthy state. From the formation of the chest, it did not appear that Corder would have been a likely subject for pulmonary affection. The medical students heard demonstrations upon the respective parts that were anatomized; there were some Italian artists there, who took two or three excellent casts of his head (Fig. 5.2), as also a celebrated craniologist, who informed us that the organs of 'destructiveness and secretiveness' were strongly developed, as also that of 'Philoprogenitiveness' (or love of children); but there was a total want of every other. His forehead was flat and not disproportioned; though small, not being more than five feet six inches high, yet he was well formed, and showed a considerable share of muscle.[60]

Dissection, then, was a much more involved surgical procedure that, like anatomisation, could serve the needs of a demanding audience.

However, the decision of how much to cut the body—either for the public audience or in the more private contexts of medical teaching and research—was still constrained by one key factor: time.

Corpses had an extremely limited shelf-life as useful anatomical objects before they were claimed by decay and putrification. In the period of the Murder Act, embalming and preservation techniques were rudimentary at best, and as discussed, medical men were frequently faced with difficult decisions on how long they should wait to ensure a body was dead, knowing that there was a finite amount of time for their work. Further, the timing of the Assizes (which coincided with the summer

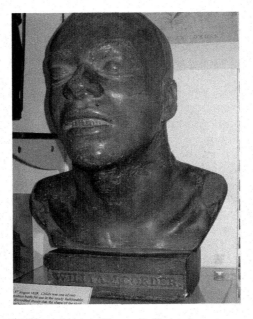

Fig. 5.2 Bust of William Corder (Photo: S. Tarlow)

sessions) meant that outside London, one of the two annual sessions at which murderers were tried, convicted, and sentenced to execution and post-mortem punishment occurred during warm months—higher temperatures accelerated decomposition, further reducing the time during which the body would be useful and safe for dissection. As Hurren has noted, it was not often possible to perform all of the anatomical procedures surgeons might wish to execute on a single body before it became overly decomposed. Instead, a 'key skill [of penal surgeons] was to dissect the maximum amount as the biological clock ticked'.[61] Medical men were highly motivated, therefore, to execute their public duties under the Murder Act and satisfy the crowd as quickly and efficiently as possible, so that they could move toward the dissection and the anatomical procedures that had professional or medical value.

There were instances in which the sentence of anatomisation and dissection was not carried out or was only partially accomplished. Earl Ferrers was the only peer of the realm convicted and punished under the Murder Act, and there was considerable interest in the question of how much his body should be cut by the medical men. In the end, his body was anatomised and the surgeons made 'a large incision from the neck to the bottom of the thorax or breast, and another across the throat', then opening the abdomen and removing the bowels'.[62] After the body was exposed to public view (both before and after the body was cut), it was dissected no further and it was taken away and buried quietly (Fig. 5.3). In the case of Thomas Gordon, hanged in August 1788 for the murder of a Constable Linnell in Northamptonshire, he had attracted considerable public sympathy by the time of his execution. As a result, 'the surgeons', the newspapers reported, 'with great humanity gave up the body to the old man and the hearse brought it to the inn where

Earl Ferrers, as he lay in his Coffin at Surgeons hall.

Fig. 5.3   Earl Ferrers in his coffin (Wellcome Collection)

Mr. Gordon resides when at this town'.[63] A few Murder Act corpses were spared the full extent of their post-mortem punishment because in some rural areas there was a lack of available and willing medical men to conduct the anatomisation and dissection.[64] But these cases are outliers. The requirements of criminal justice, medical research and teaching, and the need to satisfy the execution crowd combined and meant that nearly all mandated post-mortem punishments were indeed carried out.

Though constrained by the terms of their official role, public expectations, social hierarchies, and the limited period during which bodies remained useful for anatomical work, medical men nonetheless exercised a degree of discretion.[65] Thus, as Peter King has observed, 'justice was remade from the margins of eighteenth century society',[66] not in a central location in London, but in the provinces where a penal surgeon's reputation was made, remade and sometimes broken, in ways to which we now turn.

ACCESS AND AMBITION

By condemning individuals convicted of murder to the additional post-mortem punishment of dissection, the Murder Act created an unprecedented level of legally sanctioned and secure access to human corpses for surgeons, physicians and anatomists in Britain. This is the most compelling reason why medical men cooperated with the criminal justice system, even though the association could be detrimental to their public reputation. The service they performed in the public eye offered at best slight benefits in terms of experience or research. Much more important for the penal surgeon was ensuring effective and rapid access to bodies after their duties in the name of justice were complete. Access via the Murder Act was safer and faster than through any other mechanism at the time. Alternative methods for obtaining corpses, such as grave robbing or purchasing corpses from corrupt sextons or undertakers, may have been poor second choices in terms of safety and security, but they were by far the main source of the increasingly large numbers of bodies required during a time when direct experience of working on and learning from the dissection of human corpses was gaining importance in medical training. In order for a surgeon to build a reputation that would attract fee-paying students and patients, secure access to bodies safe to use in highly public ways was required. In practice, this is what the Murder Act provided. This section considers how access to bodies and to corpses of maximum medical value worked under the Murder Act, and the ways that the medical men harnessed the power of the criminal corpse to enhance and establish their professional reputations during the rapid professionalisation of medicine in Britain.

The corpses of executed criminals had been granted to medical men for two centuries before the advent of the Murder Act, but not without controversy. Although the royal provision of four corpses a year to the Company of Barber Surgeons, and other similar arrangements (see Chapter 3), gave official sanction to medical men to *claim* selected bodies, it offered them no protection in actually taking possession of or working with corpses. Angry crowds or the families of the condemned at times challenged the medical men and their agents at the foot of the gallows as they competed for the body, and altercations could quickly become violent and unmanageable.[67] This changed under the Murder Act: attempting to rescue a condemned murderer was made punishable by death; trying to remove a corpse from the possession of the medical men and their beadles was made punishable by transportation. With the advent of the Murder Act, for the first time strict and clearly mandated punishments for interfering with the bodies of those sentenced to dissection and anatomisation served to protect the medical men.

The Act also blunted the desire of many members of the public to prevent the anatomisation and dissection of corpses. By reserving post-mortem harm at the hands of the surgeons as a punishment exclusively for those found guilty of murder, widely understood to be the most socially transgressive crime, the State reduced the likelihood that friends or family would try to rescue these particular bodies. Even if the horror of the crime was in doubt or did not sever familial ties of affection and loyalty, family and friends would have to contest with the disapproval of the crowd—once likely to turn on medical men, now often ardent defenders of their practice—and their potential disappointment at being denied the spectacle of anatomical post-mortem punishment. Legally and socially, the Act made the bodies of murderers sentenced to dissection and anatomisation safer and more easily accessible to the medical men than other bodies they might pursue or receive—either from the gallows or the grave.

'Good Bodies': Damage, Decay and Timing

Improved access involved not only the ability to safely secure a murderer's corpse from the site of execution, but also the speed with which the body could be made ready for anatomisation and dissection. Prior to the Murder Act, when medical men were reliant on sourcing corpses (except the few allocated through royal decree or other grant) through purchase, arrangement with the hangman or gaoler, or from fresh graves to meet the needs of teaching and research, securing a body shortly after death, while it was still warm or simply before decay set in, was unlikely.

The conditions under which medical men claimed bodies under the Murder Act from the gallows significantly decreased the time delay between execution and anatomisation. The Act's removal of impediments and provision of state assistance in moving and securing the body meant that the medical men could begin work on corpses much sooner after death—and indeed, were required to do so to satisfy the post-execution crowd. One of the key implications of this shift in the timing of anatomical work is that the bodies secured under the Murder Act were likely to be safer and possibly of better quality for meeting the needs of surgeons and anatomists.

Aside from the risk of being chased or attacked by an angry mob, working with corpses regardless of their criminality was dangerous. Embalming, in the form of arterial injection, was still being developed and was a rare and imprecise practice in the period of the Murder Act. No form of refrigeration (other than standard cool cellars) was available to slow the growth of both natural and invasive organisms after death.[68] Sluicing the corpse with cold water, both before and after the penal surgeon started his work, was a popular way of improving the safety and longevity of corpses by washing away putrefying material and effluvia, removing some bacteria and microorganisms, and cooling the body. Similarly, shaving in preparation for anatomical work helped to make the body easier to work with and reduced the chance of transmission of some diseases by removing the habitat for lice and fleas. All the same, the risk of infection from diseased bodies carrying communicable diseases or parasitic infections was a hazard for medical men in the eighteenth and nineteenth centuries. Just a small nick from a scalpel could mean death from poisoning in the age before antibiotics and the regular use of antiseptics. William Rowley noted in 1795 that the anatomist must 'risk his own life to be serviceable to others'.[69] In this sense, the 'safety' of bodies vis-à-vis their anatomical use was an issue no matter their provenance. However, some risks—particularly those associated with purification—increased as the interval between death and dissection grew. Under the Act, then, timing was a key factor in the ability of the medical men to access bodies in the most useful state possible.

Hurren has written about the importance medical men placed on obtaining 'good bodies' during this period.[70] In this sense 'good' bodies were those most useful to anatomical teaching and research. In contrast, 'bad bodies' were dirty, decayed and damaged, as well as those contaminated such as by disease or lice. Freshness of the corpse, as discussed above, was a critical feature of a 'good' body, but so was the condition of the body when it reached the medical men. A body that had been roughly handled on the gallows might be much less useful for the purpose of anatomical study. If the condemned resisted the noose, or if the hangman had to make extra efforts to ensure death on the gallows by, say, hanging a second time or pulling on the body to bring about death, the corpse was likely to be damaged. Organs that had been mashed or mangled or bones broken (all possible outcomes of execution, especially if prisoners resisted) disrupted the possibility of treating the corpse as a useful, anonymised and generalised anatomical object from which conclusions could be applied broadly to the living or compared with other dead. In this context, it is no surprise that Hurren has identified the willingness of medical men to use the lancet to ensure or cause the death of those hanged under the Murder Act as a strategy often intended to reduce rough handling of the body and thereby improve its use as an anatomical specimen.[71] No matter how much some might perceive this as illegal or immoral interference, medical men could be quite practical in protecting their own interests.

Finally, murderers convicted under the terms of the Act could be considered 'good' bodies based on their treatment while waiting to be hanged: convicted murderers were allowed only bread and water in the interval between sentencing and execution. Though the intent of this clause was to serve the interests of the criminal justice system (as discussed in the last chapter), in practice there was a clear benefit to the medical men. The contents of the stomach of a corpse could give off such a foul stench that people were occasionally knocked out if that organ was nicked during dissection, and the contents of the stomach could pollute or corrupt the surrounding body before the penal surgeons had finished their work. However, in the descriptions of dissections of bodies obtained under the Murder Act, usually very little was found in the stomach as a direct result of the punitive and restrictive pre-execution diet. This diet made the bodies safer, cleaner, and easier to work with. Further, the medical men were able to compare the anatomy of bodies that had been exposed to the same food and drink as each other—an important consideration when we consider contemporary understandings of how a variety of consumption practices, from binge drinking to eating highly acidic or fatty foods, can dramatically affect the body even over short periods of time. This level of standardisation of the research object (the body of the condemned) was valued by the medical men, and was only possible because of the rigidity of pre-execution treatment prescribed by the Murder Act.

Corpses in the Countryside: Changing Patterns of Distribution of Anatomical Subjects

The Murder Act changed more than the *quality* of bodies legally available to the medical men—it also changed the *quantity* of bodies nationally available for anatomical work through clearly mandated legal means and shifted the geography of scientific access to fresh cadavers.

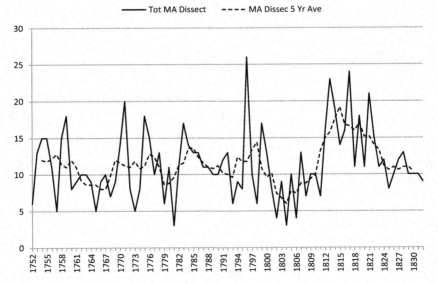

Corpses made Available to the Surgeons under the Murder Act, 1752-1832
(including Admiralty cases)

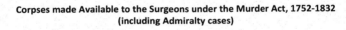

Fig. 5.4 Annual supply of bodies to medical men during the period of the Murder Act

A total of 1150 individuals were convicted and sentenced under the Murder Act over its eighty-year 'life' (Fig. 5.4). Of these, 908 were subjected to the post-mortem punishment of dissection. The average number of corpses available annually to the medical men, 11, certainly represents an increase over the few bodies previously permitted to the Colleges each year in the period preceding the Act under royal allowances and conventions (see Chapter 3). But the increase is even smaller than it might appear. There was no consistency in the number of bodies available, and while fluctuations meant that some years saw significant numbers of corpses reaching the medical men through the courts, other years saw only a very slight increase. It remained nationally true, however, that the vast majority of cadavers needed for education and research were acquired through other, extra-legal or even illegal channels.

There are broad trends in the number of legally available bodies. Most notable is the decline toward the end of the period. One of the most powerful narratives about crime and punishment in the eighteenth and nineteenth centuries is that the period evidences a transition from punishment of the body to 'gentler' forms of punishment designed to reform individuals. This period overlaps with that famously studied by Foucault, who argued that systems of criminal justice had moved away from public spectacles such as breaking on the wheel, to private discipline and punishment by removal from society, giving rise to prisons, asylums and similar penal institutions.[72] However, though the number of convictions for murder under the Act decreased over our eighty-year period, this was no smooth or linear process (Fig. 5.5). The noticeable peaks suggest that greater nuance is required when tracing social change in these centuries. Further, there was a clear push from both legislators and a variety of social commentators throughout the eighteenth century to increase the number of crimes for which post-mortem punishment would be used, but these were never successful (see Chapter 4). Regardless, it is a mistake to talk about increases or decreases in the availability of corpses for dissection under the Murder Act period without attending to wider sociopolitical contexts.

These contexts also include the regionalisation of England, and the increasingly stark divisions between 'the city' (usually meaning London) and 'the countryside'. To date, London has attracted a disproportionate amount of attention in studies of British medicine and anatomy, and while this interest is somewhat justified by the availability and centralisation of judicial and medical material history, the dominance of

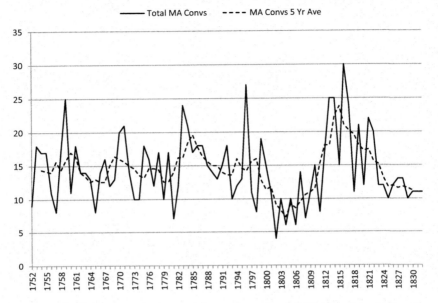

Fig. 5.5 Number of convictions per annum under the Murder Act, during the period of its operation

London over the country's medical practices changed sharply in the mid-eighteenth century. One of the key factors in London's dominance was that until 1752, London surgeons—the members of the Company of Barber-Surgeons—were the only English medical men permitted legal access to corpses for anatomical work. This created a bottleneck for anatomical teaching and research, reinforced by the monopoly of the Companies on recognised status and the lack of established hospitals elsewhere in the country that could serve as teaching and research hubs. With the advent of the Murder Act, the spatial availability of legal corpses for anatomical work shifted, spreading across England as any centre where the assizes were held (and murderers tried, convicted, sentenced and executed) became a potential source of legally sanctioned bodies for penal surgeons. This, coupled with the professionalisation of medicine and the rise of voluntary hospitals in burgeoning centres of industrial Britain over the eighteenth century, contributed to challenging London's anatomical and medical supremacy.

Even as the spatial distribution of legally available corpses spread beyond the capital, however, it did not necessarily answer to the demands of the medical men but rather to the needs of the judiciary. Medical men may have been able to access bodies outside of London, but they still depended on state structures to produce, protect and move those bodies.

As discussed in the previous chapter, there are a number of important geographical factors to consider in understanding how and why the Murder Act was applied and enforced more in some areas than others. Just as the locations of trials, executions and public displays of anatomised criminal corpses varied considerably, so too did the geographic distribution of the bodies available to surgeons under the Murder Act.

Under the Murder Act, we have found that 148 criminal corpses were produced and anatomically consumed in London over the life of the Act—only 16.3% of the total. Prior to the Murder Act, London (and more specifically, the Barber-Surgeons) had been allocated four criminal corpses a year for anatomical purposes. The (on average) two bodies per year available to London penal surgeons under the Act represents a 50% increase to the city's legal allocation of corpses for anatomical work. However, the shift in availability was nowhere near as significant in London as it was in regions outside the metropolis. For the counties, our project has demonstrated a far more striking change. Before 1752 there was no formal allowance of bodies to surgeons beyond London, Oxford and Cambridge; afterwards, all regions of England experienced an increase. By 1804, 'a penal surgeon had a much better chance of dissecting on a regular basis from legal sources that became available in the provinces rather than the capital'.[73] In fact, the legal provision of bodies under the Murder Act helped to radically decentralise London's importance to anatomy and medicine by the first half of the nineteenth century. In the last three decades of the Act's life, anatomical research was not only possible outside London in regional centres but the conditions under which bodies were received were often more favourable, resulting in new and important research on the human body.

The Value of the Criminal Corpse

The corpses that the medical men were able to access legally under the Murder Act and through their participation in the criminal justice system, however, represented only a fraction of all the bodies actually used in anatomical research and teaching. Though better documented for the Victorian period, grave robbing to supply medical centres with bodies remained common and a much-needed source of anatomical material, even during the period of the Murder Act.[74] Beadles employed by the Companies and anatomists themselves made arrangements to purchase or otherwise procure the bodies of convicts executed for other capital crimes from the gallows which, while not always strictly illegal, was difficult and lacked guarantees. Medical men may have to fight the friends and family of the deceased, asserting their 'natural rights' to procure the body of their deceased loved one. Alternatively, the gaoler or executioner might 'sell' the body to one surgeon prior to execution, then turn around and sell the body again to one or more other surgeons, leaving the medical men to fight amongst themselves. And while the selling of bodies in this way was often not strictly illegal, it was not precisely legal either and often raised public ire, meaning that there were few means by which medical men could seek redress for 'bad deals'.

It remains difficult to estimate the number of non-Murder Act bodies that entered the anatomical supply chain during the life of the Act, due to the covert nature of the transactions and resulting lack of formal paperwork. *The Diary of a Resurrectionist*, based on the diary of Joshua Naples, a body snatcher who recorded his list of activities from 1811–1812, does provide a useful picture. In London in the first decades of the nineteenth century, 'the number of subjects annually available for instruction amounted to between 450 and 500' and it was estimated that about 500 students each year were working at dissection.[75] Compare this to the approximately six bodies a year made available through legal means, and the scale of the use of corpses sourced through other means comes into focus. These bodies, though useful, were tainted by their illicit, unsavoury, or flatly illegal provenance. Anatomists were already grappling with negative public opinion due to their interest in bodies and the criminal or ghoulish practices required to procure them. To make public use of such bodies in demonstrations, lectures, or research, risked reinforcing pejorative associations to the detriment of the surgeon's reputation (and that of the profession). Worse, courting the risk of being exposed to public censure, to use bodies of questionable provenance publicly might prompt inquiries from the police, and criminal charges. For all these reasons, the criminal corpses available to medical men under the Murder Act were particularly valuable.

In the eighteenth and nineteenth centuries, medicine was not necessarily a well-paid profession. For this reason, ambition was crucial—a surgeon needed to develop a wide and dynamic reputation in order to capitalise on their skills. A surgeon had to do a good deal of self-promotion to establish the kind of reputation that would ensure a robust client list, particularly one made up of individuals and families that paid regularly. In addition, it was desirable for medical men to diversify their income streams to insulate against the vagaries of medical practice and extend their professional reputation. Teaching, public lectures or demonstrations, and original research were key activities for an ambitious—and solvent!—surgeon. And these activities required anatomical material: that is, bodies that could be used in highly visible ways that were legally 'safe'. Criminal corpses obtained under the Murder Act fit the bill. As Hurren has noted of Sir William Blizard, he was 'an ambitious man determined to establish his reputation in medical circles by undertaking gallows work so that he could stand centre-stage in the best dissection theatres of London'.[76] Indeed, 'To establish a good business reputation for medical innovation it was important to be seen to receive bodies from the hangman in a local area on a concerted basis'.[77] Surgeons worked over years to secure preferential access to the legally sanctioned bodies available under the Murder Act. Though taking on a role within the criminal justice system by anatomising and dissecting the bodies of convicted, executed murderers risked strengthening the connection in the public eye between criminality and anatomy, it was worth it if the result was the opportunity to build a reputation and profitable career based on completely above-board anatomical practice on legally sanctioned bodies.

In addition to being employed for the practice and demonstration of anatomical procedures, ambitious surgeons made use of criminal cadavers to bolster their professional reputation by engaging in original research. The results of experimentation on the bodies made available under the Murder Act carried little risk in terms of public disgust or disapproval as their crime had already given rise to social exclusion, and could therefore be disseminated and demonstrated because the bodies were legally obtained. Elizabeth Hurren has identified the types of research conducted on Murder Act corpses by 1800, and it varied widely as a result of the variability in availability of these criminal corpses.[78] For example, research on the brain and medical death took place in Leicester, on the brain and nervous system in Derby, and on gonorrhoea, heart resuscitation and breast tissue in Ipswich. A particularly intriguing area of research undertaken in the first decades of the nineteenth century involved the application of electricity to a recently dead body. Galvanism took its name from the Italian scientist Luigi

Aloisio Galvani (1737–1798) who discovered 'animal electricity' when he found that the muscles of dead frogs twitched on the application of an electrical spark.[79] Proponents of galvanism saw the best possibilities in experimentation on the bodies of the recently dead, but not those who had died of a disease. Again, Murder Act corpses fit the bill. The application of electricity to a recently dead body provoked muscle contractions that could make a corpse twitch and jerk, and in one particularly arresting demonstration on the corpse of Matthew Clydesdale (d. 1818), after connecting rods to the diaphragm and the left phrenic nerve, his chest rose and fell as if still breathing—a scene that horrified witnesses.[80] That the evidently dead corpse exhibited movement mimicking life was a source of fascination and horror (Fig. 5.6). The possibility that galvanism could lead to the reanimation of a corpse was a subject of discussion and astonishment, and is mentioned by Mary Shelley as an influence on *Frankenstein*.[81]

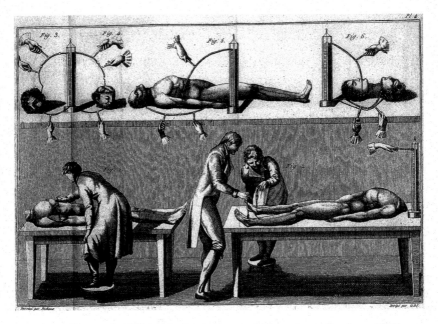

Fig. 5.6 Giovanni Aldini conducting experiments in galvanism (Wellcome Collection)

The End of Dissection and Anatomisation in the Criminal Justice System

In 1832 the anatomisation and dissection of convicted murders was removed as a judicial punishment in Britain. The reasons for this were many, including changing public opinion and the long decline in executions in Britain more generally, but the needs of the medical men played a significant role. First, the Burking scandal of 1828 in Edinburgh in which a series of 16 murders were undertaken by William Burke and William Hare in order to sell the corpses to anatomists, underscored the urgent need to reform the system of body supply for medical training and experimentation. There simply were not enough bodies made legally available to medical professionals, and while the Murder Act had made some bodies of better quality accessible to penal surgeons, the shortfall was severe and access still limited within the wider medical profession. Reflecting changing public opinion and in an effort to make more bodies available to medical men, the Anatomy Act was passed in 1832. It removed anatomisation and dissection as judicial punishments, and made the bodies of paupers unable to cover the costs of burial available to be claimed for medical training and research. However—like some bull-necked prisoners—the connection between dissection and criminality died hard, and persisted in the minds of many for some time. Just as convicted murderers balked at the prospect of dissection and anatomisation as a post-mortem punishment, so too did the poor and vulnerable display clear horror at the prospect of their bodies being cut, interfered with, or kept from decent burial.[82]

Today, dissection remains a core aspect of medical training. It is increasingly common for people to donate their bodies voluntarily for scientific and medical research. That dissection remains an important part of medical training, and social attitudes towards the practice have shifted in broad terms, underscores how the continued advances in medical knowledge today are built on the foundations established in the eighteenth century. The medical men who conducted dissections under the Murder Act were clearly trying to balance multiple needs and expectations—of the courts, of the crowds, of fellow medical practitioners and while it is difficult to track specific anatomical 'discoveries' through the penal dissection of executed murderers in England, it is safe to say that these dissections at the least ensured that understanding of human anatomy spread and grew, among both medical professionals and the general public.

More than this, however, the involvement of medical men in the dissection of criminal corpses showed the need to professionalise and standardise the conditions under which bodies were secured and new medical professionals trained. By bringing the work of surgeons into the public eye, the scandals of body snatching (and Burking) were balanced against a narrative of dissections as increasingly 'normal' and beneficial, as the celebrated philosopher and social reformer Jeremy Bentham advocated.[83] Further, just as the specific references to 'surgeons', 'anatomisation' and 'dissection' in the Murder Act led to a significant change in how medical men approached working with bodies, so too did the Anatomy Act seek to separate 'dissection' from criminality.

But not all criminal corpses could be co-opted into a narrative of progressive human betterment. A substantial minority of those convicted under the Murder Act were destined to participate in an entirely different spectacle of the macabre: hanging in chains.

NOTES

1. See for example, Gatrell, V.A.C. (1996), *The Hanging Tree: Execution and the English People 1770–1868* (Oxford: Oxford University Press).
2. See, Bennett, R.E. (2018), *Capital Punishment and the Criminal Corpse in Scotland 1740–1834* (Palgrave Macmillan); Hurren, E.T. (2016), *Dissecting the Criminal Corpse: Staging Post-execution Punishment in Early Modern England* (London: Palgrave Macmillan).
3. See for example, McCorristine, S. ed. (2017), *When Is Death?: Interdisciplinary Perspectives on Death and Its Timings* (Palgrave Macmillan); Tarlow, S. (2017), *The Golden and Ghoulish Age of the Gibbet in Britain* (London: Palgrave Macmillan); Hurren, E. (2016), *Dissecting the Criminal Corpse: Staging Post-execution Punishment in Early Modern England* (London: Palgrave Macmillan); Davies, O. and Matteoni, F. (2016), *Executing Magic: The Power of Criminal Bodies* (Basingstoke: Palgrave).
4. 25 Geo II c.37. An Act for Better Preventing the Horrid Crime of Murder (British law).
5. See, Hurren, E. (2016), *Dissecting the Criminal Corpse: Staging Post-execution Punishment in Early Modern England* (London: Palgrave Macmillan).
6. In addition to Gatrell's seminal. *The Hanging Tree*, the work of Simon Devereaux is particularly useful on this subject.

7. See, Hurren, E.T. (2016), *Dissecting the Criminal Corpse: Staging Post-execution Punishment in Early Modern England* (London: Palgrave Macmillan), quote at p. 5.

8. 25 Geo II c.37. An Act for Better Preventing the Horrid Crime of Murder (British law).

9. Hurren, E.T. (2016), *Dissecting the Criminal Corpse: Staging Post-execution Punishment in Early Modern England* (London: Palgrave Macmillan).

10. As Sean McConville has observed, 'Regular prison employment of surgeons did not begin until after 1774, when the Gaol Distemper Act permitted magistrates to appoint 'an experienced Surgeon or Apothecary' to attend the prison'. See, McConville, S. (1981), *A History of English Prison Administration* (London: Routledge), quote at p. 76.

11. See, Hurren, E.T. (2016), *Dissecting the Criminal Corpse: Staging Post-execution Punishment in Early Modern England* (London: Palgrave Macmillan), p. 3.

12. See, Richardson, R. (2000), *Death, Dissection and the Destitute* (Chicago: Chicago University Press).

13. 25 Geo II c.37. An Act for Better Preventing the Horrid Crime of Murder (British law).

14. Always conscious, of course, of the role of the crowd in influencing these decisions, discussed briefly in the previous chapter and expanded further in this chapter.

15. See, Hurren, E.T. (2016), *Dissecting the Criminal Corpse: Staging Post-execution Punishment in Early Modern England* (London: Palgrave Macmillan), p. 71.

16. See discussion of the medical and osteological process of hanging in, Waldron, T. (1996), 'Legalised Trauma', *International Journal of Osteoarchaeology*, Vol. 6, Issue 1, 114–118.

17. The 'long drop' was introduced in the 1880s, though, as noted by Vic Gatrell, it did not always guarantee a quick death. See, Gatrell, V.A.C. (1996), *The Hanging Tree: Execution and the English People 1770–1868* (Oxford: Oxford University Press), pp. 53–54.

18. For a comparative view of European execution methods, see Davies, O. and Matteoni, F. (2016), *Executing Magic: The Power of Criminal Bodies* (Basingstoke: Palgrave).

19. For a discussion of the effects of short drop hanging on the body, see Gatrell, V.A.C. (1996), *The Hanging Tree: Execution and the English People 1770–1868* (Oxford: Oxford University Press), pp. 45–46; Hurren, E.T. (2016), *Dissecting the Criminal Corpse: Staging Post-execution Punishment in Early Modern England* (London, Palgrave Macmillan), p. 75.

20. See, Hurren, E.T. (2016), *Dissecting the Criminal Corpse: Staging Post-execution Punishment in Early Modern England* (London: Palgrave Macmillan), p. 75.
21. Ibid.
22. Ibid.
23. Ibid., p. 34.
24. For accounts of executioners hanging on the legs of the body to ensure death, see Gatrell, V.A.C. (1996), *The Hanging Tree: Execution and the English People 1770–1868* (Oxford: Oxford University Press), p. 48.
25. Elizabeth Hurren describes the infamous case of 'half-hanged MacDonald', in *Dissecting the Criminal Corpse* (Chapter 1, she quotes the *Newcastle Courant*, 14 October 1754). Ewan MacDonald was convicted of murder in September 1754 in Newcastle, and sentenced to anatomisation and dissection. After his body was cut down from the gallows and laid out on the dissection table, he revived and sat up. A young surgeon promptly dispatched him with a mallet.
26. This data is available online at http://www.capitalpunishmentuk.org/half_hanged.html (Accessed 19 July 2017).
27. The number and geographical distribution of reprints attest to the popularity of Winslow's work. See for example, Winslow, J.B. (1746), *The Uncertainty of the Signs of Death, and the Dangers of Precipitate Internments and Dissections, Demonstrated* (London: Printed by M. Cooper), reprinted in 1748 in Dublin by George Faulkner; Winslow, J.B. (1749), *Dissertation sur L'incertitude des signes de la Mort et L'abus des enterremens & embaumemens précipités*, trans. J.J. Bruhier (Paris: De Bure l'Aîné), reprinted in 1752; more recently in 2010 by Gale ECCO, Print Editions.
28. Winslow, J.B. (1746), *The Uncertainty of the Signs of Death, and the Dangers of Precipitate Internments and Dissections, Demonstrated* (London: Printed by M. Cooper), p. 9.
29. Ibid., quoted at p. 23.
30. Ibid., p. 24, quoting, Terilli, D. (1615), *De causis mortis repentinae tractatio in qua etiam disputatur quid sit mors et vita in genere et quae mortis causae communes, singula vero quae de causis mortis repentinae enarrantur* (Venice).
31. Ibid., p. 23, quoting Riolanus or John Riolan (the Younger).
32. The murder is described in *The Newgate Calendar*, popular in the eighteenth and nineteenth centuries as a collection of stories associated with executions. An online version is available at http://www.exclassics.com/newgate/ngintro.htm (Accessed 21 April 2017).
33. See, Hurren, E.T. (2016), *Dissecting the Criminal Corpse: Staging Post-execution Punishment in Early Modern England* (London: Palgrave Macmillan), p. 91.

34. Ibid., quote on p. 52.
35. Ibid., p. 59.
36. Ibid., p. 96.
37. Ibid., p. 43.
38. On the crowd, see Gatrell, V.A.C. (1996), *The Hanging Tree: Execution and the English People 1770–1868* (Oxford: Oxford University Press); Laqueur, T. (1989), 'Crowds, Carnival and the State in English Executions, 1604–1868', in Beier, A.L., Cannadine, D., and Rosenheim, J.M. (eds.), *The First Modern Society: Essays in English History in Honour of Lawrence Stone* (Cambridge: Cambridge University Press), pp. 305–355.
39. Laqueur, T. (1989), 'Crowds, Carnival and the State in English Executions, 1604–1868', in Beier, A.L., Cannadine, D., and Rosenheim, J.M. (eds.), *The First Modern Society: Essays in English History in Honour of Lawrence Stone* (Cambridge: Cambridge University Press), pp. 305–355, quote at p. 309.
40. See, Linebaugh, P. (1975), 'The Tyburn Riot Against the Surgeons', in Hay, D., Linebaugh, P., and Thompson, E. (eds.), *Albion's Fatal Tree: Crime and Society in Eighteenth-Century England* (New York: Pantheon Books), pp. 65–119.
41. This extends to the other post-mortem punishment mandated in the Act, hanging in chains, as will be seen in the next chapter.
42. For example, Mary Hindes, who was sentenced to execution and hanging in 1768 for the murder of a child was 'moved considerably' by the dissection part of her sentence, according to the *London Gazetteer*, 1 July 1768. In 1785, William Higson was described in the *General Evening Post* of Saturday, 9 April 1785 as being more shocked at the dissection part of his sentence than at the death itself.
43. See for example the multiple interpretations of the reason for William Jobling's execution and gibbeting, in the next chapter.
44. See, Hurren, E.T. (2016), *Dissecting the Criminal Corpse: Staging Post-execution Punishment in Early Modern England* (London: Palgrave Macmillan), p. 199.
45. Ibid., p. 174.
46. Ibid., p. 182.
47. Ibid., p. 184.
48. Ibid.
49. Ibid., quote at p. 210.
50. Ibid., p. 209.
51. Ibid.
52. See, Hurren, E.T. (2016), *Dissecting the Criminal Corpse: Staging Post-execution Punishment in Early Modern England* (London: Palgrave Macmillan).

53. Hurren noted that in Surgeon's Hall (London), in the mid-eighteenth century, repair bills indicate that glass had to be replaced from time to time because the crowd sometimes stoned the building to protest 'a controversial criminal dissection'. See, Hurren, E.T. (2016), *Dissecting the Criminal Corpse: Staging Post-execution Punishment in Early Modern England* (London: Palgrave Macmillan), quote at p. 35.
54. See, Hurren, E.T. (2016), *Dissecting the Criminal Corpse: Staging Post-execution Punishment in Early Modern England* (London: Palgrave Macmillan).
55. Ibid., p. 41.
56. For a comprehensive account of this infamous case and further detail about his punishment, see McCorristine, S. (2014), *William Corder and the Red Barn Murder* (Basingstoke: Palgrave Macmillan).
57. *The Standard*, Wednesday 13 August 1828, Issue 387.
58. Shaving served several purposes: it removed the habitat for certain parasites therefore making the body safer for the medical men to handle—see, Hurren, E.T. (2016), *Dissecting the Criminal Corpse: Staging Post-execution Punishment in Early Modern England* (London: Palgrave Macmillan), p. 92—and it exposed the skull which made the popular dissection procedure of craniotomy easier and more visually effective, allowed phrenologists to make their examinations, and also permitted the taking of plaster casts for future research or display.
59. The seven anatomical methods of a 'complete dissection' are listed in, Hurren, E.T. (2016), *Dissecting the Criminal Corpse: Staging Post-execution Punishment in Early Modern England* (London: Palgrave Macmillan), p. 161. They are listed as follows: Osteology (study of bones); Sarcology (study of the soft or fleshy parts of the body); Myology (study of the muscular system); Splanchnology (study of viscera and its vital organs situated in the thoracic, abdominal and pelvic cavities of the body, primarily heart and lungs, but also intestines and kidneys); Angeiology (study of the circulatory system and the lymphatic system, including arteries, veins and lymphatic vases); Neurology (study of the brain and nervous system); Adenology (study of the glands and hormonal system).
60. *The Standard*, Wednesday 13 August 1828, Issue 387.
61. See, Hurren, E.T. (2016), *Dissecting the Criminal Corpse: Staging Post-execution Punishment in Early Modern England* (London: Palgrave Macmillan), quote at p. 161.
62. See, *Memoirs of the life of Laurence Earl Ferrers, Viscount Tamworth. Together with a more particular and circumstantial account of his Lordship's behaviour during the whole time of his confinement, and at the place of his execution, than has hitherto been published* (1760) (London: Printed for J. Coote, at the King's-Arms, in Pater-noster-Row), p. 56.

63. For newspaper reports see, Oracle Bell's New World, 21 August 1789; London Chronicle, 20 August 1789; London Gazette, 7 April 1752; London Evening Post, 21 March 1752. For an excellent and extensive investigation into this case, see Gray, D. and King, P. (2013), 'The Killing of Constable Linnell: The Impact of Xenophobia and of Elite Connections in Eighteenth-Century Justice', Family & Community History, Vol. 16, Issue 1, 3–31.

64. For more detail see, Hurren, E.T. (2016), Dissecting the Criminal Corpse: Staging Post-execution Punishment in Early Modern England (London: Palgrave Macmillan).

65. Ibid.

66. See, King, P. (2006), Crime and Law in England, 1750–1840: Remaking Justice from the Margins (Cambridge: Cambridge University Press).

67. See, Linebaugh, P. (1975), 'The Tyburn Riot Against the Surgeons', in Hay, D., Linebaugh, P., and Thompson, E. (eds.), Albion's Fatal Tree: Crime and Society in Eighteenth-Century England (New York: Pantheon Books), pp. 65–119.

68. During the eighteenth century, technical developments in the science of embalming the dead allowed for the preservation of teaching and research specimens, but such processes were limited in scope and occurrence through most of the period considered in this book. See, Mayer, R. (2000), Embalming: History, Theory and Practice (New York: McGraw Hill).

69. See, Rowley, W. (1795), On the Absolute Necessity of Encouraging, Instead of Preventing or Embarrassing the Study of Anatomy... Addressed to the Legislature of Great Britain (London: ECCO), p. 6.

70. See, Hurren, E.T. (2016), Dissecting the Criminal Corpse: Staging Post-execution Punishment in Early Modern England (London: Palgrave Macmillan), pp. 70–71.

71. Ibid., p. 36, pp. 56–59.

72. Foucault, M. (1977), Discipline and Punish: The Birth of the Prison (New York: Pantheon Books).

73. Ibid., quote at p. 178.

74. See, Hurren, E.T. (2012), Dying for Victorian Medicine: English Anatomy and Its Trade in the Dead Poor, c.1834–1929 (Basingstoke: Palgrave Macmillan).

75. See, Bailey, J.B. (1896), The Diary of a Resurrectionist (London: S. Sonnenschien & Co.), quote at p. 70.

76. See, Hurren, E.T. (2016), Dissecting the Criminal Corpse: Staging Post-execution Punishment in Early Modern England (London: Palgrave Macmillan), quote at p. 155.

77. Ibid., quote at p. 126.

78. Ibid., p. 248.
79. See, Lawrance, R.M. (1853), *On the Application and Effect of Electricity and Galvanism in the Treatment of Cancerous, Nervous, Rheumatic, and Other Affections* (London: H. Renshaw).
80. See, Bennett, R.E. (2018), *Capital Punishment and the Criminal Corpse in Scotland 1740–1834* (Palgrave Macmillan).
81. Mary Shelley, Preface to the 1831 edition, *Frankenstein*.
82. On the impact of the Anatomy Act, see the excellent work of Richardson, Ruth (2000), *Death, Dissection and the Destitute* (Chicago: Chicago University Press).
83. Bentham gave his own body for dissection when he died in 1832 to help reduce the stigma of dissection and to encourage others to voluntarily donate their bodies for scientific research.

CHAPTER 6

Hanging in Chains

A total of 144 individuals were executed then gibbeted in Britain under the Murder Act (1752–1832). Also known as 'hanging in chains', gibbeting was a spectacular post-mortem punishment whose impact far exceeded the relatively small number of criminal corpses that were suspended between earth and sky to be displayed for days, weeks, months, years and even decades. Sarah Tarlow has written comprehensively on *The Golden and Ghoulish Age of the Gibbet in Britain*, including the use of this form of post-mortem punishment before the advent of the Murder Act and its use by the Admiralty.[1] Here we take a more specific temporal and judicial focus, using the Murder Act to frame our examination of the formalised use of the gibbet by civil authorities in Britain from 1752–1834.[2] We begin with a description of gibbeting in Britain as a practice and process involving specific technologies, people and places. We then move on to tell three gibbet stories: William Jobling in Jarrow (1832), Spence Broughton in Sheffield (1792), and Marie-Josephte Corriveau, who was gibbeted overseas under British law (1763). The stories of the punishment of Jobling, Broughton and la Corriveau (as she is known) under the Murder Act told in this context and constructed through historical sources, archaeological evidence, and narratives, reveal complex social perceptions of what it meant to be considered 'criminal', and how the bodies of criminals were treated in relation to the interests of the state. The experiences of these three condemned prisoners, and the crowds, officials, and other actors involved in their execution and post-mortem punishments, complicate straightforward 'common sense'

© The Author(s) 2018
S. Tarlow and E. Battell Lowman, *Harnessing the Power of the Criminal Corpse*, Palgrave Historical Studies in the Criminal Corpse and its Afterlife, https://doi.org/10.1007/978-3-319-77908-9_6

narratives of deterrence and justice in eighteenth- and nineteenth-century crime and punishment, as we set out in Chapter 1. Finally, this chapter considers the legacy of the gibbet in Britain, including the power this punishment and these criminal corpses held, but also the ways in which the gibbet was part of globalising processes of carcerality and punishment through its use in Britain's overseas holdings during the period of the Murder Act.

The last gibbeting in Britain took place in the summer of 1832, after the passage of the Anatomy Act appeared to some judges to leave hanging in chains as the only available option for murder convictions. Before 1832, the gibbet had largely fallen out of use in nineteenth-century Britain. Following a public outcry, it was taken off the books in 1834. Though nearly 200 years have since passed, representations of hanging in chains arise often in Britain and North America. Whether in popular film and television or Halloween decorations, gibbets seem to be more common in the imagination of entertainment media than they ever were in real life. Media portrayals of gibbeting can be found in several major motion pictures, such as the cage in which Robin Hood's father was punished and died in *Robin Hood: Prince of Thieves* (1991), or the pirate skeletons Captain Jack Sparrow passes swinging in the wind during the opening scene of *Pirates of the Caribbean: The Curse of the Black Pearl* (2003).[3] They are also common in literature, whether in various nonfiction representations of Tudor history, or in lighthearted fantasy novels—gibbets even exist in Terry Pratchett's *Discworld* where they follow the form used and serve similar purposes as in eighteenth- and nineteenth-century Britain.[4] Museums and attractions where original and replica gibbets are displayed, including the medieval Guildhall in Leicester continue to attract those with a curiosity for ghoulish local histories nearly two centuries after the practice was abolished in Britain (Fig. 6.1). Undeniably, the gibbet is still with us, and continues to loom large in popular imagination.

Our contemporary beliefs and the historical realities of the gibbet are not always aligned, however. Under the Murder Act, gibbeting became a much more complex practice than these various later representations normally portray, in terms of both the legal procedures leading to the punishment, and the requirements of the physical process. Involving purpose-built structures for the suspension and display of hanged criminals, gibbeting was more art than science. The eighty-year period of the Act never saw the emergence of a clear consensus on best practices, either in the judicial realm or in the material matter of constructing the gibbet itself. First and foremost, contrary to some pop-culture portrayals—including the popular 1988 fantasy film *Willow*, in which Madmartigan

Fig. 6.1 Replica of James Cook's gibbet cage of 1832, now in Leicester guildhall (Sarah Tarlow)

first appears on screen imprisoned in an iron cage suspended in the air—only the dead were hung in chains in Britain in the eighteenth and nineteenth centuries.[5] In accordance with the Act, convicted criminals were first hanged by the neck until dead, and then their lifeless body was suspended on the gibbet. Of the 144 individuals gibbeted under the Act in Britain, we have identified no women, and all evidence seems to confirm that in this period only men were gibbeted.[6]

As a punishment, like post-mortem dissection and anatomisation discussed in Chapter 5, gibbeting was intended to inspire terror among witnesses and onlookers. It involved suspending the corpse of a convicted murderer between earth and sky, thereby exiling the criminal body to a liminal space, and leaving it there for up to several decades until there was little, if anything, left. For the condemned, sentencing made them aware that their body would be denied proper burial, and would be exposed, subject to public scorn, and would visibly decay, drop and be devoured by animals and insects. The criminal body might be further subjected to the ignominy of being stolen or carried off—at times, piece by piece—as decay allowed bones to fall through the gibbet cage onto the ground. In other cases, decay left an assemblage of bony body parts

Fig. 6.2 John Breads's
skull survives within the
cage of his gibbet at Rye
(Sarah Tarlow)

in the cage from which they could not be easily extracted (especially skulls, the only bones which were unlikely to fit through the cage without assistance)[7] (Fig. 6.2). Certainly there was no peaceful 'rest' to anticipate that might ease a troubled mind before execution.

Gibbeting was also intended as a deterrent to the commission of heinous crimes by others: it was expected to, and often did, inspire horror, terror and revulsion in onlookers through the denial of funeral rites and desecration of the corpse. The humiliating display of the body, its eerie and uncanny motion on the gibbet, and the disgusting smells and excretions emanating from the corpse as it decomposed, all contributed to this spectacular, arresting punishment. And insofar as the decaying bodies of gibbeted criminals served to enforce the law through fear, the gibbet was also a key factor in asserting state power through social horror.

These generalities aside, however, the historical life of the gibbet is diverse and complex. We speak of 'the gibbet' as if it was a single, straightforward object, but the term is actually simple shorthand for a complex nexus of techniques and technologies.[8] That is to say, there was a great deal of variation in how the material gibbet was produced and used, and these differences are important for the kinds of social discourses that sprang up around the occasion and location of gibbeting, whether those in support of state power, in sympathy for the condemned men, or otherwise. Making sense of the range of gibbet techniques and technologies employed under the Murder Act requires extensive comparison which is enabled by the compilation by the Criminal Corpse team of a list tracing as many surviving material remains of and textual allusions to gibbets as possible.

WOOD, METAL, LAND AND FLESH: MAKING GIBBETS

There are three types of evidence that, taken together, have made it possible for us to undertake an entirely unprecedented engagement with one of the most evocative but understudied forms of punishment in British history.[9] First, sheriffs in eighteenth- and nineteenth-century Britain were responsible for organising the construction and erection of the gibbet, engaging guards to provide security at the gibbet site, and overseeing the corpse's progress from gallows to gibbet. The Sheriffs' Cravings are the detailed reimbursement claims for the costs sheriffs incurred in the course of carrying out this punishment.[10] Investigated for the Criminal Corpse project by historian Richard Ward,[11] this hitherto underused source lists in fine detail the materials and services required to gibbet a man under the Murder Act. Second, archaeologist Sarah Tarlow conducted a comprehensive survey of all existing gibbet cages (the part that encases the body) in Britain today, and was able to identify 16 whole or partial cages, allowing for unprecedented comparative analysis.[12] Third, work with textual sources including newspapers, pamphlets, broadsheets, ballads and images provided valuable information on the spectacle and sociopolitical role of the gibbet in Britain. By combining information from these sources, it has been possible to construct an accurate idea of the gibbet as built and used by civil authorities during the life of the Murder Act.[13]

All British gibbets share common features and yet are also unique. They are perhaps best understood as variations on a core theme, in which a number of different factors played into the creation of the individual forms. We have identified six distinct elements that comprise all gibbets used during our period of study: a corpse, metal cage, hook and/ or short chain, crossbeam, post and erection site. Each of these features can vary in any number of ways while still fulfilling their function as part of the gibbet assemblage. We turn here to a close examination of each element, beginning with the erection site.

Fig. 6.3 (continued)

Fig. 6.3 Gibbet locations. **a** St. Peter's rock, Derbyshire, Anthony Lingard's gibbet took advantage of this natural landmark. **b** A gibbet sited on the riverbank in Sleaford took advantage of its visibility from both the road and the river. **c** Abraham Tull and William Hawkins were gibbeted near this crossroads in Berkshire (all photographs: Sarah Tarlow)

Place is important to the function of gibbets, and spatial considerations of the gibbet are a key element of this technology. Gibbet sites were carefully selected and often the sentencing judge indicated a general space or site in the sentence, sometimes one relating to the crime committed by the condemned prisoner. Many were erected at or near the location of the crime, the trial, or other aspect of the judicial proceedings.[14]

However, the gibbet could not be located at the scene of the crime without consideration of other, more practical concerns (Fig. 6.3). Key to site selection was the need for both security and accessibility. Like hangings, gibbetings and the gibbet itself drew large crowds. It was not unusual for thousands, and sometimes tens of thousands, of people to visit new gibbets, and they remained sites of interest and visitation so long as the gibbet stood. Limited access combined with large crowds risked property damage, riots, and other dangers of large-scale disorder. In addition to allowing sufficient space to accommodate the curious and carnivalesque crowd, the site needed to help ensure the visibility of the gibbet. High places such as hills or locations next to major travel routes or at crossroads were all sites that accomplished this goal. Locations such as on commons, wasteland or open heath also allowed large crowds to gather safely.[15] A final consideration in selecting a gibbet site was managing the sensory experience of the gibbet, by which we mean that a body in various states of decay gives off a variety of pungent and deeply unpleasant smells while also being distinctly visually unappealing. Officials were sometimes petitioned to relocate gibbets located near the homes or properties of prominent individuals, requiring additional work and expense. The gibbets of Abraham Tull and William Hawkins in Berkshire were taken down and buried at the request of a well-to-do local lady. William Andrews recorded that 'Mrs. Brocas, of Beaurepaire, then residing at Wokefield Park, gave private orders for them to be taken down in the night and buried, which was accordingly done. During her daily drives she passed the gibbeted men and the sight greatly distressed her.'[16] Gibbets positioned near family homes, such as the one at Lower Hambleton, now submerged beneath Rutland Water, contributed to enduring guilt and infamy being attached to not just the convict, but also their surviving kin.[17] In that case the parents of the two brothers suspended there could see their sons' remains from their front door.

Gibbet locations were chosen both because of the existing physical landscape, and also because of the social and cultural landscape that became entangled with these unmistakable, grisly objects. Their semi-permanence, the attention they drew from local people and visitors, and the stories of the crimes of the gibbeted individual (see Chapter 8) transformed these sites and changed their relationships to nearby places and the people who passed through—just as the site impacted the longevity and social impact of the gibbet.

Moving from site to structure, gibbets were erected by first securing a sturdy post in the ground. Thus, tethered to and drawing strength and stability from the earth, gibbets were constructed to stand for years, decades, or longer. The gibbet required a thick wooden beam, about 10 metres long that was planted firmly in the ground—in at least one case, secured there with a foundation stone—so that it stood upright.[18] The post had to be strong enough to support the weight of the crossbeam and heavy iron gibbet cage year-round for an indefinite period. The height of the post had to be great enough to make the swinging body visible from a distance. The height also helped to complicate potential efforts to rescue (or steal) the corpse, and some posts were made more secure by studding them with thousands of nails, or plating the lower part of the post in iron to make it very difficult to saw through (Fig. 6.4).

The next element of the gibbet was a wooden crossbeam (Fig. 6.5). This beam was attached near to or at the top of the gibbet post at a 90-degree angle. It had to be of sufficient length to allow the gibbet cage to be attached to the end opposite the post without resting against it. Here also, the crossbeam had to be strong enough to support the weight of the gibbet cage and body outdoors for up to several decades.

An iron hook and/or short length of chain were the means by which the cage was attached to the crossbeam and suspended in the air.

Fig. 6.4 Post of a gibbet (possibly Parr's) with nails used to reinforce it and make it harder to saw through, now in Banbury Museum (Sarah Tarlow)

Fig. 6.5 The crossbeam of a gibbet allowed the caged body to swing freely (Sarah Tarlow)

The hook and chain, like the post and crossbeam, had to be sturdy enough to support the weight of the cage and corpse long term. However, a critical aspect of this element of the gibbet is that, while it firmly connected the cage to the wooden structure, at the same time the hook or chain had to permit the cage to move—that is, to swing or rotate freely. The gibbet cage of John Breads (executed in 1743 for the murder of his brother-in-law, Allen Grebell) held in the Rye museum, Sussex, shows the sort of device that was used to attach the cage to the crossbeam. The metal of the top of the hook is visibly worn down on the Rye gibbet, demonstrating not only that movement of the suspended cage occurred, but that the cage experienced significant and regular movement during the more than 20 years that Breads' gibbet stood in Gibbet Marsh (Fig. 6.6). Movement played a key role in the way that the public experienced gibbets. As the wind caused the gibbet to sway, the metal-on-metal sound of the hook and chain caused an eerie noise that was especially unsettling at night.[19]

Fig. 6.6 Hook of
Breads's gibbet, Rye,
showing wear (Sarah
Tarlow)

The cage is the most visually arresting element of the gibbet. Though
the punishment was also known as hanging in chains, the gibbets used by
civil authorities during the life of the Murder Act did not wrap the body
in chains, but rather encased it in a purpose-built cage made of inflexible
iron bands. These held the body securely and the cage was attached via
the hook and/or chain to the crossbeam. Even empty, its shape continued—and continues—to evoke the body it once held (or would hold).
There are a variety of gibbet cages on record, most of which were shaped
to both contain the body of the criminal, and to evoke the human form
even after the body decayed. Tarlow has found from her work with the
16 extant examples in Britain, that there is considerable variation in the
form and construction of gibbet cages, and no local design traditions can
be observed.[20] However, the similarities between these artefacts reveal
important components in their construction and use under the Murder
Act. First, the cage of the gibbets held its occupant in an upright (standing) position. Though this could be accomplished with supports under
the crotch and sometimes under the feet, and an articulated head piece
to hold the body in this position, a key factor in accomplishing this is

Fig. 6.7 Keal's gibbet, Louth Museum, only encloses the head and torso (Sarah Tarlow)

that cages were invariably suspended from the crossbeam by attaching the hook and/or chain to a point at the top of the head. Second, it held the body securely so that it could not be easily removed or slip free. As is visible in Keal's gibbet at Louth, Lincolnshire, the bands were cinched tight around the torso and other large body parts (Fig. 6.7). Cages invariably encased the torso and head, though some also had bands to secure the arms and legs. That the cage had to fit tightly to the body in order to constrain it is what gave rise to its anthropomorphic form (Fig. 6.8). Third, the cage allowed physical and visual access to the body. The iron bands could not be so thick as to completely obscure the corpse. Rather, the cage had to permit the visibility and recognisability of the body. This aspect of the gibbet also allowed animals, birds and insects access to the swinging corpse, encouraging decay and disintegration of the body. The cages used as part of punishment under the Murder Act were single-use items and there is no evidence to suggest they were reused. Rather, like the other parts of the gibbet, the cage was intended to be sturdy enough to secure the body and last for decades exposed to the air and elements.

Fig. 6.8 Punched holes on the gibbet allowed adjustment to fit the body of the criminal (Sarah Tarlow)

The standing position and anthropomorphic shape of the cages contributed to two important and eerie effects of the gibbet. First, that the body in the gibbet so clearly dead—since this was an expressly post-mortem punishment and local people were very likely to have witnessed the execution—seemed somehow and unnervingly alive. The standing position, the swaying, swinging, or turning in the wind, all created an uncanny and paradoxical impression of 'life'. Further, as a body decayed, was consumed, or pieces dropped to the ground, the cage might become empty, bit by bit. The absence of the body, however, in the continuing presence of the swinging, turning, man-shaped cage only reinforced the unsettling nature of the punishment.

The final element of the gibbet is the corpse, which is literally and figuratively at the centre of this form of post-mortem punishment. As mentioned, civil authorities (unlike the Admiralty courts) did not reuse gibbets, which created a one-to-one relationship between a particular corpse and its gibbet. This meant a gibbet could be referred to by the name of the criminal it held or by its location with ease and specificity. The corpse was brought by cart from the place of execution, which might be near to or identical with the selected gibbet site, as discussed above. A good example of this is the case of John Walford, who was gibbeted in 1789 at the scene of his execution in the Quantock Hills of Somerset. In contrast, criminals executed by the Admiralty might be transported some distance from the site of execution in London to the site of their gibbet in, say, Devon or Norfolk. Some secondary literature makes reference to covering the corpse in tar or pitch,[21] as in the case of Tom Otter whose body was said to have been covered in a layer of pitch before being reclothed, presumably to aid with identification, and enclosed in the gibbet cage. However, there is no evidence of such a practice in the Sheriffs' Cravings even though small and inexpensive items, such as ale for the guards or rope for a noose, are frequently listed. Further, as the cage was designed to be form fitting, a thick layer of pitch might have caused difficulties for properly placing and securing the corpse. It is unclear, in any case, what the purpose of the pitch might be since there was clearly no desire to try and preserve the body. Rather, once the body was sealed into the cage, it was unlikely ever to be removed from it except in the course of progressive decay—a key element to the horror of the gibbet intended to deter future crimes.

PROGRESS AND PUNISHMENT: DID THE GIBBET WORK?

In addition to the material technique of gibbeting, involving the physical elements (place, post, crossbeam, hook and chain, cage and corpse), the gibbet was and remains a product of discrete but connected discourses of punishment. For the state, gibbeting during the period of the Murder Act was intended to serve two complementary purposes. First, as Elizabeth Hurren has argued, and as was discussed in Chapters 4 and 5, the Murder Act constituted an expression of the common law principle *lex talionis*. Derived from Roman law, *lex talionis*—captured succinctly by the adage 'an eye for an eye'—refers to the reestablishment of social order following a severe transgression by inflicting upon the malefactor the harm they visited on their victim. The punishment for murder under the Act therefore stripped the murderer of their own life and debased the body as murder did to the victim, in this case by denying proper burial. Gibbeting was intended to accomplish this goal in a way both similar to and different from public dissection. The medical men might dissect a body to the point where it could no longer be recognised as an integrated individual, but suspension in the gibbet obliterated the body as a social object, exiling it to the liminal space between earth and sky, and did so publicly in order that the communities affected by the crime could witness that justice had been done. Second, the state intended the spectacular punishment of those who had transgressed one of the most fundamental human social laws (the unsanctioned killing of other humans) to act as a deterrent to further commission of the crime by others. The humiliating, powerless exposure and display of one's body, and the knowledge of how the impending punishment affected the condemned before death—including stories of 'hard men' such as Lambert Reading in 1775 who were unfazed by the idea of execution but could not hear the sentence of gibbeting with equanimity—were intended as an instructive lesson to the many who witnessed, read about, or spoke of the punishment.[22] For the State, gibbeting was intended to restore social and State cohesion and function following a transgression, while simultaneously reducing the probability of future incidences of the crime being committed by others. For all the effort accorded these aims, how far were they actually achieved at the foot of the gibbet, or in the pages where the punishment was represented for an eager public? There is no unambiguous answer.

There were no clear guidelines within the Act regarding which post-mortem punishment a judge selected when sentencing a murderer. Both were equal in the eyes of the law. However, it is possible to discern in practice one convention and two probable factors that directed judges in their decision to choose the gibbet over anatomisation and dissection. First, as mentioned, of the 144 instances of gibbeting under the Murder Act in Britain identified by our research, we know that no women were gibbeted. They were 'invariably sent for anatomical dissection.'[23] The reason why female murderers were always sentenced to dissection and anatomisation rather than gibbeting in the eighteenth and nineteenth centuries is less clear. We do know that female bodies were highly sought-after by the surgeons and that this added value or demand may have swayed the decision to dissect rather than hang in chains.[24] Further, as Peter King has noted, social sensitivities in Britain regarding the treatment of the bodies of executed female criminals may have extended to a general and strong reluctance to gibbet women.[25] In early modern France, such chivalrous concern for decorum sometimes led to the refusal to execute women by hanging, preferring instead to bury them alive.[26] However, as we will see, the British aversion to gibbeting women evaporated when we travelled into the overseas British world.

Beyond the issue of sex, age and availability may have played a pragmatic role in guiding judges to choose gibbeting when sentencing convicts. In cases where more than one murderer was sentenced at the same time and location, a judge might decide to split the fate of the condemned, sending some to the surgeons and some to the gibbet. In 1784 father and son John and Nathan Nicholls were both convicted of and executed for the same crime in Suffolk. The son's body was taken by the surgeons, but the older man was hung in chains, perhaps because his body was a less desirable anatomical object. A decision of this kind may have been rooted in several considerations: if surgeons in an area were unable to make use of multiple bodies (particularly at times of the year that promoted rapid decomposition) it may have been unfeasible to send the body to surgeons elsewhere, undesirable because it removed the instructive opportunity of the post-mortem punishment from the community most affected by the crime, or risky if it deprived the gallows crowd from witnessing the conclusion of the punishment spectacle. Surgeons probably made clear their practical preferences when multiple bodies were potentially available to them under the Act. Robust, young bodies were in demand for medical purposes and those of the aged or infirm, much less so. Gibbeting was much more expensive per body than anatomisation and dissection and was much more spatially intensive—the display of the anatomised body required a suitable room for a day or two, but the gibbet required an open plot of land for decades. Differential post-mortem treatments of murderers' bodies did not follow any hard and fast rules, however. Rather, judges took into consideration numerous logistical and judicial concerns when deciding which post-mortem punishment should be accorded a convicted murderer.

The ignominy of the long-term public display of the corpse as it decomposed on the gibbet, and its relative rarity compared to anatomisation and dissection, might suggest that this punishment was reserved for those who committed the most violent and shocking murders (Fig. 6.9). Certainly, beyond the Murder Act, civil authorities reserved their discretionary use of the gibbet to criminals sentenced for particular types of crimes. Those that threatened the authority of the state for all manner of treasons have a history of being subjected to post-mortem punishment, particularly the quartering of criminal corpses but also display on

Corpses Hung in Chains under the Murder Act, 1752-1832 (including Admiralty cases)

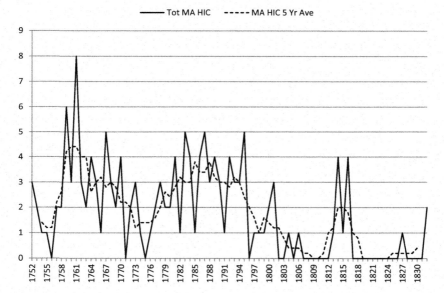

Fig. 6.9 The frequency of gibbeting from before the Murder Act to its end (Sarah Tarlow)

the gibbet.[27] Robbing the post and highway robbery were also by custom punished in this way, as they were seen as crimes that disrupted the natural flow of people, information, and money—mobilities that served the interests of the state, even if indirectly.[28] However, under the Murder Act there is no evidence to suggest that judges used any particular metric of the horror of a murder to determine post-mortem punishment.[29] Judges worked within their understandings of the law, the sociopolitical context in which a crime and its prosecution occurred, and the parameters of practical considerations such as capacity, cost, time and impact.

And yet, gibbeting under the Murder Act seems out-of-step with prevailing understandings of the changing nature of punishment during this period. The late eighteenth century saw the rise of confinement as a specific form of punishment along with discourses of rehabilitation rather than the *lex talionis* approach centred on punishment and retribution. Before this, accused and convicted criminals were held in gaol awaiting either trial or execution of their sentence (i.e., the gallows, transportation) but only as a temporary measure. However, the interruption and discontinuation of transportation to the American Colonies in 1775, following the American Revolution, forced British authorities to rethink their punishment practices. Confinement and hard labour, first on the prison hulks moored in ports across Britain, then later in large purpose-built prisons, alongside the resumption of transportation (this time to Australia and with a more distinct colonising purpose), rapidly took root.[30] Those involved with engineering the shift away from capital punishment in Britain towards confinement and the penitentiary explained their efforts as a move away from the barbarity of corporal and capital punishment.[31] However, in the second half of the twentieth century, historians and sociologists re-engaged with the transition from the gallows to the penitentiary and questioned the benevolent, civilising nature of this process. Powerful explanations such as the need to meet the growing demands of the labour market, the use of confinement as a way to re-establish social control in the context of class-based unrest, and as a method of more effectively exerting sociopolitical power and control through regimes of social discipline, were all posited as driving forces.[32] Most potent, however, is the narrative of the civilising process. As Elias proposed, 'civilization' is marked by 'a reduction in the use of physical violence and an increase in the intensity of psychological control.'[33] The display of rotting corpses in prominent public places certainly seems to have no place in this civilising trajectory. So why did the practice continue after this shift?

In fact, gibbeting in general and more specifically under the Murder Act did decline in frequency from the end of the eighteenth century. As Tarlow has found, gibbeting by civil authorities peaked *before* the punishment was formalised in law when the Act went into effect in 1752.[34] By 1800, gibbeting was such a rare occurrence (excluding in the Admiralty courts) that only 10 took place under the Act between 1800 and 1834 when the punishment was legislated out of existence in Britain. There were probably multiple factors in this decline, including the cost, space needed, and ever-increasing need of bodies for surgical practice and training. In the early nineteenth century, we do know that gibbeting offended the sensibilities of onlookers and communities, and that the public expression of repugnance at the continuing existence of the punishment became increasingly insistent. When the punishment was repealed in 1834, the legislation passed rapidly and with little discussion.

This was a result both of the strong public aversion to such grisly punishments and of the fact the Murder Act had effectively been made redundant by the passing of the Anatomy Act in 1832 (see Chapter 4).

But the opposition to gibbeting in the nineteenth century is also a testament to its puissance. The gibbet transformed a body into something dreadful with power over places, stories and people. The spectacular punishment impacted individuals, families, communities and nations. Just as each gibbet was unique, so too is each story.

Three Gibbet Stories

William Jobling, 1832

He committed no murder, on that point both victim and accused agreed. The hand that struck the ultimately fatal blow was not that of William Jobling, but his involvement in the mortal attack on Nicholas Fairles on 11 June 1832 was never in question. The *degree* of his involvement was contested at trial, but under the law failure to decisively act to prevent murder carried the penalty of death on the gallows. So, in the eyes of the law at least, Jobling's execution seems to have been inevitable. He was hanged by the neck until dead on 3 August 1832 by executioner William Curry outside Durham Gaol.[35]

But that was not the end of Jobling's punishment. The Durham Assizes judge, Mr. Justice Parke, in accord with what he believed to be the requirements of the Murder Act, passed sentence of death on the gallows but also directed that the corpse be hung in chains. Nearly two hundred years later, Jobling's name is far from forgotten; not only for his actions in life but because of the manner of its ending and the fate of his corpse (Fig. 6.10).

To say that tensions were running high in the coalfields in northeast England in the spring of 1832 would be putting it mildly. The Great Strike of 1831 may have ended in victory for the miners and improvement in the terms of their yearly contracts ('bindings'), but the unions formed in the course of that conflict that were so critical to the strike's success became the object of ire for mine owners by the following year.[36] In 1832, mine owners objected to the growing strength of the unions and refused to 'bind' any man who was a member. New strikes began in March, and things escalated quickly. Families were evicted from their cottages, and by spring, more than eight thousand pitmen were on

Fig. 6.10 Replica of
Jobling's gibbet, South
Shields Museum (Sarah
Tarlow)

strike. Nearly the same number had returned to work and were supporting those on strike with pay contributions. As David Ridley has argued,
this situation was more than a labour dispute: industrial unrest, a major
cholera epidemic, and the parliamentary reform campaign of 1831–1832
made for a critical and acute state of crisis. There were assaults, riots, and
a simmering unrest that threatened to boil over at any point.[37] For this
reason, Nicholas Fairles, Esq., a Magistrate of the county of Durham
took up temporary residence at the Jarrow Colliery to be on hand to
prevent further breaches of the peace. Fairles was, by all accounts, a
well-respected elder member of the community. As Alan Marshall discovered, Fairles was energetically involved in upholding the law: he
once ordered the seizure of 500 cakes and rolls having found them to
be deficient in weight, and gave the confiscated bread to the poor. On
another occasion, he intervened to prevent a medical man from procur-
ing corpses from the Constable and the churchyard.[38] Stern, smallish,
septuagenarian Fairles was returning on a pony on the afternoon of 11
June 1832 when he met William Jobling, just past a turnpike on the road
from Jarrow Colliery.

On the day everything changed, William Jobling, pitman, husband to Isabella (née Turner), and father to several young children, was drinking. For a 30-year-old man who had likely begun working in the mines before the age of 10, this pursuit was something of a given. Jobling had been out of work since 5 April, and now some two months later, lacked the coin to pay his way. His solution was to linger by the road and ask passers-by to 'treat him with a quart of ale'.[39] He met with some success because we know he was given a shilling by John Arthur Foster (of the Jarrow colliery) for this purpose. When Fairles rode by Turner's public house at about five o'clock, Jobling approached him, laid a hand on that of Fairles, and with good grace asked for money for a drink. Fairles refused, noting that the man—who was known to him—had appeared to have already had 'a sufficiency'.[40] At this point, another man came up behind Fairles, took hold of his coat and dragged him from his horse.

Eyewitnesses saw two men setting upon the Magistrate and all three struggling on the ground. 'One of the men rested on Mr. Fairles, and struck him with a large stick, and the other held him down', and Mary Taylor and her aunt, Margaret Hall said they heard one of the men say 'kill him, kill him'.[41] Taylor shouted at the men to be off, and the two assailants ran away. Fairles, badly injured and bleeding, was led away by the women to a nearby house.

His injuries were severe, but Fairles lived for another 10 days. Before he expired on 21 June, Fairles gave a statement on the attack. He named Jobling as the man who had held him down, and Ralph Armstrong, a pitman of Jarrow colliery, as the man who had attacked him from behind, and battered his head with stones and the heavy, horn stick Fairles was accustomed to carry. It didn't take long to locate Jobling, a man witnesses who spoke in his defence called quiet, harmless and inoffensive, 'a notorious coward' unlikely to engage in such an aggressive, violent act.[42] Armstrong, on the other hand, never stood trial. He absconded after the attack, and at the time of Jobling's trial, Armstrong was still at large—despite hundreds of pounds offered as a reward for his apprehension.[43] This pitman who had also been in the employ of the Jarrow Colliery, was 'about 44 Years of Age, 5 feet 9 inches high, stout made, Dark Complexion, Blue Eyes, large Mouth, large turned-up Nose and Brown Hair',[44] and was never caught.

But for Jobling, there was no escape. Tried for murder during the Durham Assizes on 1 August 1832, Jobling's indictment charged that he was present and assisted Armstrong in murdering Fairles.[45] He pleaded not guilty. Witnesses were called, and testified to seeing Jobling stop Fairles' horse, to seeing him struggling on the ground with Armstrong and Fairles, and that both men had been seen running along the road leading to South Shields (where Jobling was apprehended), Armstrong with blood on his hands.[46] The deposition made by Fairles himself was produced and stated that Jobling had held Fairles down with hands and knees while Armstrong struck the victim's head with stones. When Mr. Cobb, police officer for South Shields, produced Fairles' stick with blood visible on the end, the newspaper reported that 'Jobling changed colour'.[47] Jobling maintained his innocence, stating that he ran away when Armstrong pulled Fairles off his horse. The jury didn't take long to consult, and in just fifteen minutes returned their verdict: guilty.

Sentence was duly passed by Mr. Justice Parke. He warned Jobling to expect no reprieve. Parke attributed the 'want of moral principle' which allowed Jobling to stand by while a man was viciously attacked to the unions—the 'combinations'—which had been active in the labour disputes between pitmen and mine owners in the region.[48] Parke called combinations 'injurious to the public interest and to those who are concerned in them' and in sentencing Jobling to a shameful death on the gallows, 'hoped to God it would be a warning to others'.[49]

But he didn't stop there. Aware that a new bill before parliament to discontinue dissection as a punishment for convicted murderers had probably already received Royal Assent, Parke kept to what remained of the Murder Act and sentenced Jobling to hang in chains and 'hoped the sight of it would have a due effect on the prisoner's companions'.[50] In accordance with the dictates of the Murder Act, the execution was scheduled for just two days later. On the platform, as he was about to be launched, 'a person near the scaffold cried out, "Farewell, Jobling", and he instantly turned his head in the direction whence the sound proceeded, which displaced the cord, and consequently protracted his sufferings, which continued some minutes'.[51] The gallows then the gibbet, erected on Jarrow Slake within sight of Jobling's wife's cottage, were guarded by soldiers. However, as soon as the guard was withdrawn, Jobling's remains were removed under cover of night. His corpse had swung in its cage for only about a month. There had been no order to bring down the body, as there was for James Cook (gibbeted shortly after Jobling in Leicester) whose gibbet was brought down by order of the Home Office after just three days.[52] Whoever rescued Jobling's corpse from the gibbet on the night of 7 September 1832 put themselves at great risk: the crime of taking a body from a gibbet still carried a penalty of 7 years transportation. However, it seemed clear to those in the area that Jobling's fellow pitmen were unhappy with the harsh punishment meted out to their compatriot, and acted thus in 'service to his memory'.[53]

And this memory lives on.

Today, songs, stories, poems and memorials tell us that Jobling was a man punished by the powerful as an example to his compatriots at the Jarrow collieries. His death and post-mortem punishment were intended as a shocking deterrent to labour organising in the nineteenth century. Jobling and his gibbet can be consumed, metaphorically, in the form of a beer called Jobling's Swinging Gibbet made by the Jarrow Brewery. Vincent Rea staged a comprehensive exhibition about Jobling and his post-mortem punishment in 1972 at the Bede Gallery that included artistic representations and a life-sized model/replica of Jobling's gibbet commissioned from artist Laurie Wheatley. The folk band The Whiskey Priests' song 'Farewell Jobling' commemorates Jobling's memory in music.[54]

The myth of William Jobling, as martyr/murderer endures, but is, of course, unlikely to provide satisfactory answers to the complicated contexts and violent outcomes of both crime and punishment in this case. Though gibbeting as a punishment was intended to remove the comfort or certainty of a final earthly resting place from the condemned, in the case of Jobling this uncertainty is magnified because though his body was rescued, it has never been found. Today, a small stone memorial stands to remind those passing of Jobling. It was erected at the former site of the Gaslight public house where the body in its cage was rumoured to have been buried after it was 'rescued' from the gibbet. But this additional absence has not deterred those drawn to his story and its enduring meaning, nor does it seem likely to.

Spence Broughton, 1792

Spence Broughton is remembered by some as the last man gibbeted in England.[55] He wasn't, not by a long shot.[56] However, Broughton *is* remembered and that is a much more remarkable feat considering that he died more than 200 years ago and achieved very little of note in his forty-six years. So why is he remembered? Primarily for what happened to his body after his death, and the years it spent on display as it mouldered, rotted and fell to pieces.

Spence Broughton was executed on the gallows outside York Castle on 14 April 1792. He had been imprisoned for 6 months before his life ended at the end of a rope. Broughton had abandoned his wife and three children to spend his time gambling, but not before financially ruining his once-comfortable family. Born in Horbling near Falkingham, Lincolnshire, Broughton was a tall, well-made man. He was reputed to be from an honest and hardworking family, but was eloquently described in a small volume devoted to his life and crime as 'a degenerated plant from a good tree'.[57] Gambling was his main pastime, and it is perhaps no surprise that he lost much more than he ever won.[58] This behaviour put his family into severe financial difficulty, and Mrs. Broughton at last secured a separation.

Broughton particularly favoured cockfighting, races, and games of chance including E O tables (a game of chance related to roulette). He lost huge sums, and when personal and family funds ran dry, Broughton turned to theft to support his dissolute habits. In the company of John Oxley and Thomas Shaw, Broughton planned and carried out robberies of mail coaches. Shaw provided information and funds, Oxley converted bills found in the mail into cash, and Broughton took the lead in carrying out the thefts.

Oxley, Shaw and Broughton robbed the Rotherham Mail carried by a boy, George Leasley, on Attercliffe Common between Sheffield and Rotherham on the night of Saturday 29 January 1791. Leasley testified that a mile-and-a-half from Rotherham, he was stopped by two men whose faces he could not see who compelled him to leave the road, 'one tied his hands and fastened him to the hedge, whilst the other cut away the bag containing the letters, with which they made off.'[59] A foreign bill of exchange worth £123 was taken from the mail, with the help of a French dictionary the bill was exchanged successfully and Oxley 'decamped with all the proceeds except for £10'.[60] Broughton went after him and after some effort obtained £40 from Oxley.[61] They then planned and executed the robbery of the Aylesbury Mail on 28 May 1791 but finding little of value, actually lost money on the venture.[62] Finally, the trio planned to rob the Cambridge Mail in June 1791. This was the most successful of their efforts. Targeting the Cambridge Mail the day after the Newmarket races meant that the mail was packed with the bank bills of London's gentleman gamblers. As before, the boy carrying the mail was taken off the road into a field and tied to a post there. The robbers buried the Cambridge letters in a nearby field, and took from the mail thousands of pounds of bank bills, nearly £5000 of which Oxley successfully cashed in London at several establishments before the robbery became public knowledge.[63]

The malefactors were apprehended when on enquiring with the bankers, the bills were found to have been stolen from the Cambridge Mail, and Oxley was identified, pursued and taken in London. He claimed the bills belonged to Shaw, and directed the thief-takers to Shaw's lodgings, but on arrival found Broughton there and after a chase, apprehended him. Shaw was not long in following. The three were held in gaol and closely questioned. Shaw turned King's evidence, laying the blame on Oxley and Broughton. Oxley escaped from gaol in October 1791, and in the end, only Broughton stood trial. He was transported to York in January 1792,[64] and at the Lent Assizes in York, 'indicted for feloniously assaulting George Leasley on the King's highway in the county of York, putting him in fear, and taking from his person the ROTHERHAM MAIL'.[65] The trial only took two-and-a-half hours to find Broughton guilty,[66] and despite his protestations that he was miles away when the crime took place Judge Buller immediately passed sentence of death.

To deter similar crimes, Judge Buller decreed that punishment should not end with execution, but rather that Broughton's body should hang in chains at the site of his crime. Executed on 14 April, the body was then taken from York to Attercliffe Common and in the small hours of Monday 16 April 1792, suspended on a gibbet near the Arrow Inn.

Certainly, Broughton was not the only highwayman gibbeted for his crimes. Though distinctly unbloody, the nature of his transgression threatened the security of property and movement of finance, which sat at the heart of the capitalist values on which the nation was built, and so warranted severe punishment in the eyes of the law. The public fascination with the image of the dashing highwayman during this period (and since) was also a risk to public order. Harsh punishment held the possibility of bursting the bubble of public approbation of such robbers. The post-mortem aspect of Broughton's punishment was specifically intended as a warning to others. For this, the grisly and highly visible nature of the suspension of Broughton's corpse on a gibbet should have been a good fit.

Instead, within hours of the erection of the gibbet post and before the body in its iron cage had been brought by cart to be hung up, the site of the gibbet of Spence Broughton attracted hundreds of people keen to see the spectacle. This is perhaps no surprise as newspapers noted the unusually high attendance at the execution.[67] The publican of the Arrow, situated near Attercliffe Common, recalled that at eight o'clock in the evening, on the day after the execution (Sunday), 'the common was like a fair'.[68] This man, George Drabble, was the one to lead the officials conveying Broughton's corpse to the site of the gibbet, some 200 yards from the Arrow. His assistance was rewarded by the incredible number of customers brought to the Arrow in the weeks and months that followed, drawn to witness the gibbet. One local recollection described the scene: 'On the first Sunday after he was gibbeted, all the road through Attercliffe was one mass of people, going to and from the gibbet. Many remarked they never saw so many people in their lives, and wondered where they came from, for it beat Sheffield fair, and seemed as if they never would give over coming.'[69] It is difficult to identify the crowd's reaction to witnessing Broughton's gibbet, but it seems to have served more as public entertainment than a solemn lesson.

Broughton's gibbet remained in place for more than 36 years. As late as 1828, it was still possible to see 'his skull, and a few bones and shreds of clothing, which had survived the storm and stress of the weather.'[70] A Sheffield local, Dr Sorby, recalled that the gibbet had stood on land belonging to his father, who was the one who took the structure down 'owing to the inconvenience of people coming to see it'.[71] That the gibbet with its mostly empty cage continued to draw curious onlookers nearly four decades after Broughton's corpse was suspended there is telling: these visitors included many not even born at the time of Broughton's execution, and for whom the gibbet itself was of much more interest than the individual for whom it had been built or his crime. In fact, in 1867 'Many hundreds of persons' came to see the excavated remains of Broughton's gibbet post[72]—a piece about four-and-a-half feet long and 18 inches square black with age—which was unearthed during excavations for new houses in Clifton Street on what was once Attercliffe Common.

Today, Broughton's story is most prominently, and fictitiously, visible in the form of the advertising for the Noose and Gibbet Inn, located near the site of Broughton's gibbet (Fig. 6.11). The Inn proclaims Broughton to be the last man hanged in England (which he clearly was not) and sports a 'replica' gibbet complete with a mannequin to greet those passing in front of the building. This is no historically accurate reproduction, the pretend irons being of the 'bird cage' variety. So what does this mean for the story of Spence Broughton and his life and crimes? Perhaps it is fitting that just as the grisly spectacle of the gibbeted

Fig. 6.11   The Noose and Gibbet Inn, Sheffield, exploits its proximity to the place where Spence Broughton was gibbeted with a wholly inauthentic recreated gibbet (Tom Maskill)

body drew visitors and served as a sort of entertainment, so today the Noose and Gibbet Inn, and its sensational representation still draw in the punters.

Marie-Josephte Corriveau, 1763

We know that no women were gibbeted in the eighteenth and nineteenth centuries in Britain. Invariably, women convicted of murder were punished with hanging and dissection or, in the event that the murder was categorised as petty treason (the killing of a male superior such as a master, father, or husband), up to the end of the eighteenth century by strangulation then by burning at the stake.[73] We remain curious as to why anatomisation and dissection, involving as it did the exposure of the opened and at least semi-nude body to public view, were somehow considered a more appropriate treatment of the bodies of female murderers than their display fully clothed in the gibbet. Whatever the reasons, the result is clear: under the Murder Act, no women were hung in chains.

Except one.

In the eighteenth century, Britain engaged in active and energetic colonisation of places and peoples around the globe. Less than two decades before the American Revolution, the British were involved in a contest with another and much more established colonial force in the eastern part of North America. In 1759, General Wolfe defeated le Général Montcalm at the Plaines d'Abraham, and ended King Louis XV of France's control of Nouvelle France.[74] This region, which today is encompassed by the Canadian province of Québec, had been settled since the early seventeenth century as a French Catholic colony. The Canadiens, as the inhabitants were known, were an agricultural people who lived according to the seigneurial system of land tenure and feudal agricultural practice. Before the British conquest, the French king kept tight controls on Nouvelle France, including prohibiting the establishment of a domestic press in order to head off the possibility of the rise or spread of rebellious ideas or actions. The English began their control of Nouvelle France following Wolfe's victory but it was not until the autumn of 1763 that civil authority was established for the province of Québec under the terms of King George III's Royal Proclamation of 7 October 1763. For the years between conquest and the advent of civil government under the British, the colony operated under military law and General James Murray served as Governor. His was the highest authority in the province and he had responsibility for confirming all sentences passed by courts martial. During this interim period, a death occurred, a court martial followed, and a woman was convicted of murder. Her name was Marie-Josephte Corriveau and her legend is inseparable from the cage in which her corpse was displayed two hundred and fifty years ago.

Marie-Josephte Corriveau was born in Saint-Vallier near Québec and baptised on 14 May 1733. She was the daughter of farmer Joseph Corriveau and Marie-Françoise Bolduc. First married in 1749 to Charles Boucher, also a farmer, she had three children before her husband died in 1760. She married again in July 1761 to another farmer, one Louis Dodier. Dodier's corpse was discovered in the early morning of 27 January 1763 in his stable. It appeared that he died at some point in the night as a result of several severe wounds to the face and head. Whether the fatal wounds had been caused by a horse, or a sharp instrument was a matter of some debate. The day before the incident, Joseph Corriveau had complained about his son-in-law during a visit to the local priest and it was said that Marie-Josephte had asked her father to beat her husband. We have access to the detail of the initial court martial, at which Joseph Corriveau was tried for murder and Marie-Josephte was tried as an accomplice, thanks to the preservation work of J.M. Lemoine who was president of the literary and historical society of Québec in the late nineteenth century.[75] The transcript of that court martial reveals a meticulous process of investigation, and includes testimony from those who had attended the scene after the body was discovered, members of the Corriveau family, people of the neighbourhood, and details from a coroner's inquest. The first trial began on 29 March 1763 and ended on 9 April. The conclusion was that Joseph Corriveau was convicted of murder and sentenced to execution, and that Marie-Josephte was found guilty of being an accomplice to the crime and sentenced to public whippings and to be branded. However, shortly afterward, Joseph made a confession. He avowed that it was actually his daughter who had committed the crime. A second court martial was convened. Joseph was proclaimed innocent, and once she had made a confession, Marie-Josephte was convicted of the murder of her husband. She was sentenced to death by hanging and to additional post-mortem punishment in line with the terms of the Murder Act.

She was hanged at Québec on 18 April 1763. Shortly after, the corpse was encased in a purpose-built cage that followed the general form of cages constructed in Britain. The body of Mary-Josephte was gibbeted for five weeks at a crossroads in St. Joseph, Point Levy, Québec before being removed and taken away for burial at a nearby churchyard, whose specific location was not made public, still encased in the gibbet irons.

In 1849, workers digging behind the church of Saint-Joseph-of-Bellechasse as part of renovation efforts discovered an iron body-shaped cage still containing a few bones. It was recognised immediately as the gibbet cage of Marie-Josephte Corriveau. The gibbet cage of 'La Corriveau', as she came to be known, became an immediate object of interest, and was sold to Barnum's Circus in New York who took it on tour to cities including Montreal, New York, and Boston where people paid to see this ghoulish curiosity. It is reported that the cage was presented by David P. Kimball to the Essex Institute in Salem, Massachusetts in 1855 where it was kept in their collections of scientific and historical materials until the early twenty-first century.[76] In 2013, however, the cage again became an object of intense interest as it was 'rediscovered' by Québécois researchers. After tests carried out on the metal of the cage and historical investigation, in 2015 'la cage de la Corriveau' was confirmed to be authentic, and was repatriated to Québec as an object of significant historical and cultural value.

But what made this artefact so special and so significant? In the years after her death and post-mortem punishment, the story of Marie-Josephte grew, twisted and transformed. As Luc Lacourcière noted in his seminal article on this history and phenomenon in 1969, scarcely a year has gone by since 1763 that this story, and the figure at the centre of it, have not been the subject of new literary and artistic representations.[77] Indeed, because of the indelible connection between woman and crime cemented by the cage that was so integral to her spectacular public post-mortem punishment, La Corriveau became one of the key figures not only in Québécois patrimoine (heritage), but in Canadian folklore more broadly. To call her story a 'legend' is by no means an overstatement. Marie-Josephte's conviction for the murder of her husband and grisly gibbeting gave rise to her being recast as an evil woman who murdered not only her second husband, but her first, and as many as five others. In the vilification that followed her punishment, she became not only a notorious murderess, but also a sorceress and a malignant spirit set on tormenting the living.

In the context of our research, Marie-Josephte is remarkable not only for what her story became, that potent legend of La Corriveau, but for the way her post-mortem punishment stands in such stark contrast to that meted out to convicted female murderers in Britain. So, why was Marie-Josephte Corriveau tried, convicted and punished in this way? That she was tried by a court martial instead of a civil process was recognised as an error by Governor Murray, who nonetheless noted that in the absence of other established structures during the interim period between formal French and British rule, he had followed a precedent set in a similar situation in Montreal two years earlier.[78] That she was sentenced to post-mortem punishment on the gibbet in addition to execution on the gallows also followed the precedent set in the 1761 case. However, if she was sentenced under British law, why was she not sentenced to burning, as would have been the appropriate punishment for a woman convicted of petty treason—and being found guilty of murdering her husband, Marie-Josephte would certainly have met the criteria for this specific crime. But if, as it seems, she was sentenced according to the Murder Act, why not anatomisation and dissection as would have been the norm had she been in Britain? It is likely few facilities existed for such a course of action, which may have impacted the decision. However, there is also the issue of when the murder and punishment took place with relation to the political situation. Both occurred before the instigation of civil rule by the British in what had been Nouvelle France, and the punishment of Canadienne (French Settler) Marie-Josephte by a British court martial suggests an effort by British authorities to use the post-mortem punishment of convicted murderers as an example to others who may have sought to instigate unrest or challenge British rule. It is possible that in her case, the analogy between treason and petty treason meant that husband-murder stood in for political insurrection, thus making such a harsh and ostentatious punishment seem appropriate.

We know that the British carried gibbeting as a post-mortem punishment as part of its colonising efforts not only to Canada, but also to America, Australia, New Zealand and India. When used to punish white British overseas subjects, gibbeting followed the form used by the civil authorities in Britain. However, the use of the gibbet to punish enslaved African people—men *and* women—particularly in the plantation colonies, was much more brutal and violent. In Antigua, six enslaved African individuals were gibbeted following an uprising in 1736. They were gibbeted alive and condemned to hang in chains to die of thirst, hunger and exposure.[79] In Jamaica, gibbeting alive was one of a suite of horrific punishments used by the white planter class to terrify and control enslaved African people.[80] Two men, Fortune and Kingston, were gibbeted alive in 1760 in Kingston, Jamaica. These enslaved African individuals were captured as ringleaders of the violent St Mary Rebellion against the white planter class and were gibbeted alive at one of the thoroughfares of the capital.[81] A sketch from that decade depicts a gruesome double gibbet on Kingston's main parade.[82] The gibbet was also used to punish enslaved black people in colonial America, such as in the case of one man 'hung alive in chains in the town' in New York City following a rebellion of enslaved African people.[83]

Yet Marie-Josephte Corriveau remains the only (white) woman we know of hung in chains under the Murder Act. For other subjects whose bodies were gibbeted in the British world, infamy is likely a more appropriate way to refer to their posthumous and in some cases, long enduring renown than the status of legend rightfully accorded to Marie-Josephte Corriveau. In Britain, ghost stories around gibbets are surprisingly rare, and certainly none approaches the recognition and notoriety of La Corriveau as she continues to be known in literature, legend and contemporary renderings. But more appropriately, we argue, that name refers to the union of a woman accused and convicted of a crime and the cage that played a key role in her post-mortem punishment. Indeed, without the gibbet, Marie-Josephte Corriveau would never have been transformed into the legendary voracious man-murderer, malignant spirit and tragic dark figure still known today.

THE GIBBET TODAY: ENDURING AND APOCRYPHAL

We are not the first, nor will we be the last to tell the stories of Jobling, Broughton, and Corriveau. Our tellings are intimately related to the context in which these stories are told here: in relationship with our focus on the criminal corpse in eighteenth- and nineteenth-century Britain. As John Lutz has said, all historical sources are partial, in both senses of the word.[84] Our tellings are similarly partial. We tell these stories here, together and assembled in this way, to move from historical summary to more individual detail, and to try to create a view into the historical use of the gibbet that may allow us to connect our own ideas, assumptions, and feelings with an event and experience that is otherwise alien.

Today, the gibbet persists in contemporary representations, artefacts and stories. Some gibbet sites are commemorated or maintained, the most well known of these being Combe Gibbet, Winter's Gibbet and Caxton Gibbet which still sport replicas of their original posts (Fig. 6.12).[85] These sites still attract the public—for example, the Combe Gibbet to Overton annual 16-mile cross-country race uses the gibbet site as a key place marker as well as the name of the event. Gibbet artefacts are popular items in the local museums that house them, and the ghoulish nature of their history appeals to a wide range of ages and interests.

The gibbet is also with us in more fanciful forms. Birdcage style gibbet Halloween decorations are sold in both the United Kingdom and North America, and similar versions can be found in various macabre entertainments, such as the popular London Dungeon experience. In these contemporary forms, the gibbet has been safely contained in museum spaces, made campy displays for macabre holidays, or has even been converted into voluntary erotic indulgence by some with a sexual taste for punishment, but in all cases the gibbet retains its entertainment value and dark fascination for those not subjected to its historical and actual ends.

Fig. 6.12 (continued)

Fig. 6.12 **a** Caxton gibbet (Sarah Tarlow) and **b** Winter's gibbet (Patrick Low). Both gibbets have been curated, restored and replaced to enable their continued function as local landmarks

NOTES

1. See, Tarlow, S. (2017), *The Golden and Ghoulish Age of the Gibbet in Britain* (Palgrave Macmillan).
2. In eighteenth- and nineteenth-century Britain, gibbeting was a punishment used by civil and military authorities, and the practice shows some variation between these two groups. The Admiralty used gibbeting to punish piracy and mutiny in a variety of traditional locations usually located along a shoreline and sometimes also reused gibbet cages/chains. Gibbeting by the civil authorities differed in terms of gibbet technology.
3. See, *Robin Hood: Prince of Thieves* (1991) [film], dir. by Kevin Reynolds (USA: Warner Bros.); *Pirates of the Caribbean: The Curse of the Black Pearl* (2003) [film], dir. by Gore Verbinski (Walt Disney Pictures).
4. Pratchett mentions gibbets in at least three Discworld novels. These include *Nightwatch* (2011), *The Fifth Elephant* (1999), and *Feet of Clay* (1996). It is worth mentioning that in the Discworld, gibbets are constructed and function in a much more historically accurate manner than

in most other fictional contexts. Their reception also follows historical example. In *Feet of Clay*, Pratchett writes: 'At the end of Nonesuch Street was a gibbet, where wrongdoers—or, at least, people found guilty of wrongdoing—had been hung to twist gently in the wind as examples of just retribution and, as the elements took their toll, basic anatomy as well. Once, parties of children were brought there by their parents to learn by dreadful example of the snares and perils that await the criminal, the outlaw and those who happen to be in the wrong place at the wrong time, and they would see the terrible wreckage creaking on its chain and listen to the stern imprecations and then usually (this being Ankh-Morpork) would say "Wow! Brilliant!" and use the corpse as a swing.'

5. See, *Willow* (1988) [film], dir. by Ron Howard (USA: Metro-Goldwyn-Mayer).

6. This is discussed further below in the context of the social discourse around gibbeting.

7. See, Tarlow, S. (2017), *The Golden and Ghoulish Age of the Gibbet in Britain* (Palgrave Macmillan).

8. This differentiation draws on the work of Ellul, J. (1954), *La Technique: L'enjeu du siècle* (Paris: Armand Collin).

9. For a discussion of the criteria by which gibbets were selected, see for example, Tarlow, S. and Dyndor, Z. (2015), 'The Landscape of the Gibbet', *Landscape History*, Vol. 36, Issue 1, 71–88; and for a consideration of the technical and design features of the gibbet cage, see, Tarlow, S. (2014), 'The Technology of the Gibbet', *International Journal of Historical Archaeology*, Vol. 18, Issue 4, 668–699.

10. The Sheriff's Cravings are records of expense claims submitted to the Treasury by each county's sheriff for the costs incurred in the punishment of all assize convicts. They can be accessed at The National Archives, London, Sheriffs' Cravings, T 64/262, T 90/148–66, Sheriffs' Assize Calendars, E 389/242–8.

11. This previously underexploited source of evidence was discovered by Richard Ward, and examined in, Ward, R. and King, P. (2015), 'Rethinking the Bloody Code in Eighteenth-Century Britain: Capital Punishment at the Centre and on the Periphery', *Past & Present*, Vol. 228, Issue 1, 159–205.

12. For a list of existing gibbet cages in Britain, see, Tarlow, S. (2014), 'The Technology of the Gibbet', *International Journal of Historical Archaeology*, Vol. 18, Issue 4, 668–699, list at page 684. We should note that not all of these cages date from the period of the Murder Act, but all were used by the civil authorities (not the Admiralty), except possibly for the one owned by Winchester Museums. The small sample size makes consideration of cages from beyond the period considered in this chapter necessary.

13. Much about the gibbet has relied on secondary literature and repeated some things as 'facts' which are otherwise unsubstantiated. This primary source investigation is unprecedented in this area.

14. See for example the case of Spence Broughton at the end of this chapter. And on the connection between crime or individual and site of punishment see Stephen Poole (2015), '"For the Benefit of Example": Processing the Condemned to the Scene of Their Crime in England, 1720–1830', in Ward, R. ed., *A Global History of Execution and the Criminal Corpse* (Basingstoke: Palgrave); Tarlow, S. and Dyndor, Z. (2015), 'The Landscape of the Gibbet', *Landscape History*, Vol. 36, Issue 1, 71–88.

15. Though many place names including the word 'gibbet' still exist today across Britain, it is worth noting that most of these relate to medieval sites of execution.

16. See, Andrews, W. (1899), *Bygone Punishments* (London: W. Andrews & Company), p. 63.

17. Sleath, S. and Ovens, R. (2007), 'Lower Hambleton in 1797', in Ovens, R. and Sleath, S. (eds.), *The Heritage of Rutland Water* (Rutland Record Series Number 5) (Oakham: Rutland Local History and Record Society), pp. 193–209.

18. A broken socket stone at Gonerby Hill Foot, Lincolnshire, is believed locally to have supported a gibbet at one time (http://www.lincstothepast.com/photograph/290331.record?pt=S).

19. See, Tarlow, S. (2017), *The Golden and Ghoulish Age of the Gibbet in Britain* (Palgrave Macmillan).

20. See, Tarlow, S. (2014), 'The Technology of the Gibbet', *International Journal of Historical Archaeology*, Vol. 18, Issue 4, 668–699.

21. One of the most well-known sources to make this claim is Hartshorne, A. (1893), *Hanging in Chains* (New York: Cassel Publishing Company).

22. *London Chronicle*, August 5–8, 1775, issue 2912.

23. See, Tarlow, S. and Dyndor, Z. (2015), 'The Landscape of the Gibbet', *Landscape History*, Vol. 36, Issue 1, 71–88, quote at p. 73.

24. See, Tarlow, S. and Dyndor, Z. (2015), 'The Landscape of the Gibbet', *Landscape History*, Vol. 36, Issue 1, 71–88.

25. As Peter King has found, no female murderers were gibbeted during the life of the Murder Act, and none of the fifty-five individuals convicted of property crimes and sentenced to hang in chains during the same period were women. King quotes noted eighteenth-century jurist William Blackstone who wrote 'the decency due to the sex forbids the exposing ... their bodies'. King goes on to point out that as women's corpses were sent for public anatomisation and dissection, the thinking behind this gendered policy remains unclear. Peter King, *Punishing the Criminal Corpse 1700–1840: Aggravated Forms of the Death Penalty in England* (Palgrave, in press), chapter 3, p. 14. Quoting William, Blackstone, *Commentaries on the Laws of England*, 2, 18th London ed.

(New York: Collins and Hannay, 1832), as quoted in Simon Devereaux, (2009, February), 'Recasting the Theatre of Execution: The Abolition of the Tyburn Ritual,' *Past & Present*, Vol. 202, pp. 127–174, p. 77.

26. See, Naish, C. (1991), *Death Comes to the Maiden: Sex and Execution 1431–1933* (London: Routledge).

27. The law at the time differentiated between 'petty treason'—betrayal of authority, as in a worker or servant betraying their employer—and 'grand treason' which is closer to the definition common today of attempting to undermine the state.

28. See, Tarlow, S. (2017), *The Golden and Ghoulish Age of the Gibbet in Britain* (Palgrave Macmillan).

29. Not that such a 'metric' is possible anyway; it's certainly not useful to or for us to try and create a hierarchy of horror against which to test a hypothesis.

30. In 1823, an act of Parliament authorised the transportation of British convicts to any colony designated by the Crown. This gave rise to transportation of English and Irish convicts in particular, to places like Bermuda where they were used for projects in support of imperial expansion. For accessible and comprehensive information on British convict transportation and the wider global context of this punishment, see the excellent outputs of Clare Anderson's *The Carceral Archipelago*, in particular her project website ConvictVoyages.org.

31. This is the narrative explored by Foucault, in *Discipline and Punish*, and Ignatieff, in *A Just Measure of Pain*. See also Markus, T. (1993), *Buildings and Power* (London: Routledge).

32. See for example, Rusc he, G. and Kirchheimer, O. (1939), *Punishment and Social Structure* (New York: Columbia University Press); Ignatieff, M. (1978), *A Just Measure of Pain: The Penitentiary in the Industrial Revolution, 1750–1850* (London: Macmillan); Foucault, M. (1977), *Discipline and Punish: The Birth of the Prison* (Harmondsworth: Penguin).

33. See, Vaughan, B. (2000), 'The Civilizing Process and the Janus-Face of Modern Punishment', *Theoretical Criminology*, Vol. 4, Issue 1, 71–91, quote at p. 74.

34. See, Tarlow, S. (2017), *The Golden and Ghoulish Age of the Gibbet in Britain* (Palgrave Macmillan).

35. See, Fielding, S. (2013), *Hanged at Durham* [ebook] (The History Press), Appendix II. Available at https://books.google.co.uk/books?id=WPo6AwAAQBAJ&pg=PT203&lpg=PT203&dq=william+jobling+hanged+date&source=bl&ots=MMb5GmM6Wg&sig=ᔓ 1VAe2NiOmRu2hPtkSgAc4byXPk0&hl=en&sa=X&ved=0ahUKE-wiThY_SpYfNAhVPF8AKHdS5DU8Q6AEIWDAL#v=onepage&q=jobling&f=false (Accessed 26 April 2017).

36. *Tyne Mercury*, 5 June, 22 and 29 May (1932), from, Ridley, D. (1994), *Political and Industrial Crisis: The Experience of the Tyne and Wear*

Pitmen, 1831–1832, Durham theses, Durham University, p. 238. Available at Durham E-Theses Online: http://etheses.dur.ac.uk/5366/ (Accessed 26 April 2017).

37. See, Ridley, D. (1994), *Political and Industrial Crisis: The Experience of the Tyne and Wear* Pitmen, 1831–1832, Durham theses, Durham University, p. 238. Available at Durham E-Theses Online: http://etheses.dur.ac.uk/5366/ (Accessed 26 April 2017).

38. Marshall, A. (2009), *The Death of Nicholas Fairles: Law and Community in South Shields, 1832* (North East England History Institute).

39. *The Morning Chronicle* (London, England), Saturday 4 August 1832, p. 4.

40. NA HO 44/29, fo. 118. Deposition of Nicholas Fairles.

41. *The Morning Chronicle* (London, England), Saturday 4 August 1832, p. 4.

42. Ibid.

43. £100 was offered by the Vestry of St. Hild's Chapel, South Shields and a further £300 was offered by the government. *The Newcastle Journal* (Newcastle-Upon-Tyne, England), Saturday 4 August 1832, Issue 13, p. 3. From database British newspapers 1600–1950.

44. *The Newcastle Journal* (Newcastle-Upon-Tyne, England), Saturday 22 September 1832, Issue 20, p. 1. From database British newspapers 1600–1950.

45. *The Morning Chronicle* (London England), Saturday 4 August 1832.

46. Ibid.

47. *The Morning Chronicle* (London, England), Saturday 4 August 1834.

48. *The Morning Chronicle* (London England), Saturday 4 August 1832.

49. Ibid.

50. Ibid., p. 4.

51. 'Execution of William Jobling for the Murder of Mr. Fairles', *Times* (London, England), 6 August 1832, p. 5.

52. Related in the *Newgate Calendar*, 'JAMES COOK, Executed 10th of August, 1832, for the Murder of Mr. Paas, whose Remains he attempted to destroy by Fire'. Available at http://www.exclassics.com/newgate/ng614.htm (Accessed 26 April 2017).

53. See, Pelham, C. (1841), *The Chronicles of Crime, or the New Newgate Calendar* (London: Printed for Thomas Tegg).

54. *Whiskey Priests* (1996), 'Farewell Jobling' [song], on Life's Tapestry [alb.] (Whiskey Priests).

55. See for example, Hindle, R. (2012), *The Purging of Spence Broughton, A Highwayman.* Available at: https://spencebroughton.wordpress.com/essays/ (Accessed 7 June 2017).

56. See, Tarlow, S. (2014), 'The Technology of the Gibbet', *International Journal of Historical Archaeology*, Vol. 18, Issue 4, 668–699.

57. See, *The Life and Trial of the Unfortunate Spence Broughton* (1792), (printed by John Drury, 3rd edition), p. 10.

58. Ibid.

59. 'News' *Leeds Intelligencer* (Leeds, England), 2 April 1792, p. 3.

60. See, *The True and Illustrated Chronicles of the Last Man Gibbeted in Yorkshire* (1900) (reprinted by January Books), p. 2.

61. Ibid.

62. *Lloyd's Evening Post* (London, England), October 17–19, 1791, Issue 5352. Public-office, Bow-street Mail-Robbers, &c.

63. Ibid.

64. 'News', *Leeds Intelligencer* (Leeds, England), 23 January 1792: 3.

65. See, *The Life and Trial of the Unfortunate Spence Broughton* (1792), (printed by John Drury, 3rd edition), p. 3.

66. 'News', *St. James's Chronicle or the British Evening Post* (London, England), March 24–27, 1792.

67. 'France', *Chester Chronicle* [Chester, England] 20 April 1792: 2.

68. Criminal Chronology of York Castle by William Knipe p. 125.

69. Hester, G. 28 January 1873, *Sheffield Independent* (Sheffield, England), 31 January 1873, Issue 5214, p. 4.

70. See, *The True and Illustrated Chronicles of the Last Man Gibbeted in Yorkshire* (1900) (reprinted by January Books), quote at p. 2.

71. *Sheffield Independent* (Sheffield, England), 4 March 1896, Issue 12931, p. 8.

72. *Stamford Mercury* (Stamford, England), 10 May 1867, Issue 8977, p. 3.

73. See, King, P. (2006), *Crime and Law in England, 1750–1840* (Cambridge: Cambridge University Press). As King explains, throughout the eighteenth century, this punishment involved death before burning. This was also a punishment meted out against female coiners. Burning of women declined as a practice in eighteenth-century Britain, and was formally abolished in 1790.

74. This battle was part of the larger French and Indian War (1754–1763) between England and France in North America, which in turn was part of the larger, global Seven Years War between those two imperial powers during the same period.

75. Bibliothèque et Archives nationales (BAnQ), Québec P1000 S3 D453, 1960-01-008/104.

76. The Essex Institute was succeeded by the Peabody Essex Museum in 1992, and the institution is considered one of the oldest continually operating museums in the United States.

77. See, Luc Lacoursiere, C.C. (1969), 'Le destin posthume de la Corriveau', *Les Cahiers des Dix*, Numéro 34, 239–271.

78. For a key article on the subject see, Luc Lacourcière C.C. (1973), 'Présence de la Corriveau', *Les Cahiers des dix*, Numéro 38, 229–264. The precedent-setting case was that of the 1761 conviction of Saint-Jean (also known as Saint-Paul) for a quadruple murder in Saint-François-de-Sales (Quebec) who was also gibbeted for his crimes. For more

information see, Trudel, M. (1999), *Le régime militaire et la disparition de la Nouvelle-France, 1759–1764* (Québec: Les Editions Fides).

79. Eighty-eight enslaved African people were put to death by the Antigua planters. Of these, 5 were broken on the wheel, 6 gibbeted alive, and 77 burned alive. See, 'America and West Indies: May 1737 16–31', in Davies, K.G. ed. (1963), *Calendar of State Papers Colonial, America and West Indies: Volume 43, 1737* (London), pp. 155–173. Available at British History Online http://www.british-history.ac.uk/cal-state-papers/colonial/america-west-indies/vol43/pp155-173 (Accessed 17 January 2017).

80. Burnard, T. (2004), *Mastery, Tyranny, and Desire: Thomas Thistlewood and His Slaves in the Anglo-Jamaican World* (Chapel Hill: University of North Carolina Press).

81. Institute of Jamaica, *The Gleaner*, 7 October, 2010. Available at http://jamaica-gleaner.com/gleaner/20101007/news/news7.html (Accessed 19 April 2017).

82. Pierre Eugène du Simitière, untitled Jamaican scene, ca. 1760s. Ink on paper, 8 × 13 in. (20 × 30 cm). The Library Company of Philadelphia. In Louis Nelson, *Architecture and Empire in Jamaica*, New Haven, Yale University Press, 2016, p. 127.

83. Letter of Governor Robert Hunter to the Lords of Trade, in O'Callaghan, E.B. ed. (1855), *Documents Relative to the Colonial History of the State of New York* (Albany, V, 1707–1733), pp. 341–342.

84. See, Lutz, J. (2008), *Maguk: A New History of Aboriginal-White Relations* (Vancouver: UBC Press).

85. See, Tarlow, S. and Dyndor, Z. (2015), 'The Landscape of the Gibbet', *Landscape History*, Vol. 36, Issue 1, 71–88.

PART III

The Legacy of the Criminal Corpse

CHAPTER 7

Seeking the Physical Remains
of the Criminal Corpse

We have traced the legal construction of the criminal corpse under the Murder Act, and its journey beyond the gallows and into the spaces and processes of post-mortem punishment. In the final section of this book, we turn our attention to the afterlives of these criminal corpses. We begin in this chapter by considering the material afterlives of bodies, partial bodies, and artefacts created from those punished under the Murder Act. Intentionally and accidentally, artefacts and objects from this period and from these bodies were preserved, and some remain with us into the present day. We ask: what kinds of physical remains endured and what are the ways that power inheres in them, then and now? In Chapter 8, we turn to the less material, though nonetheless potent and powerful, narrative remains of these punished corpses. In stories, songs, art, drama and literature, the criminal corpses created under the Murder Act linger with us and 'haunt' our everyday lives in the present. How is the power of the criminal corpse produced, and why does it still have the ability to disturb and entertain us today? Finally, in Chapter 9 we examine the philosophical and ethical legacies of the Murder Act and the treatment of corpses, criminal corpses, and criminalised corpses in Britain. For now, we turn our attention to the material remains and physical traces of the criminal corpses produced and manipulated under the Murder Act.

© The Author(s) 2018   193
S. Tarlow and E. Battell Lowman, *Harnessing the Power of the Criminal Corpse*, Palgrave Historical Studies in the Criminal Corpse and its Afterlife, https://doi.org/10.1007/978-3-319-77908-9_7

RESTING IN PEACE OR RESTING IN PIECES?

The Murder Act specifically excluded the bodies of convicted and executed murderers from burial until the corpse had been anatomised and dissected or hung in chains. However, these punishments did not have a similarly formal or legislated moment at which a body might be considered to have been punished 'enough'. There was no agreed point at which the punishment ended, and the body (or what remained of the body) could be laid to rest. If carried to their full extent, both dissection and hanging in chains disintegrated the body of the condemned, leaving very little, if anything, that would require burial or another form of disposal. As we have seen in Chapters 5 and 6, those responsible for carrying out post-mortem punishments had some discretion in the accomplishment of their duties, and in some cases could choose how extensively a body should be cut or how long a body should be left hanging in chains. In practice, the extent to which these punishments destroyed the criminal corpse was variable, as was the type and amount of human material remaining at their conclusion.

There were no formal directions for the disposal or interment of the human remains of criminal corpses created under the Murder Act.[1] At the discretion of the medical men and those involved in the maintenance of gibbet sites, the (usually partial) bodies created under the Murder Act were disposed of in a variety of ways. To better understand how and why these bodies and body parts were put to various uses after post-mortem punishment, it is useful to consider broad patterns in their disposal over the life of the Act. We begin with burial, and compare the treatment of Murder Act bodies to those later dissected under the Anatomy Act of 1832. Then we will consider three other eventual fates of the human remains of those punished under the Murder Act: their use for magical and medical purposes, for educational ends, and for the creation of curiosities including macabre souvenirs.

Criminal corpses created under the Murder Act generally did not achieve the (relatively) happy end of 'resting in peace'—that is, the burial of an intact body according to normative cultural or religious protocols. For those men convicted of murder and hung in chains, burial was extremely unlikely. But there were a few exceptions. James Cook of Leicester was the last man gibbeted in Britain. He was a 21-year-old bookbinder convicted on 8 August 1832 of the murder of Mr. John Paas, a manufacturer of brass instruments to whom Cook owed money.

The crime was discovered when neighbours saw light and smoke coming from Cook's house and found to their horror that he was attempting to burn Paas's dismembered body and destroy the evidence of his crime. Cook was hanged on 10 August 1832 outside Leicester gaol and gibbeted near the Aylestone tollgate. The execution crowd is said to have numbered about 40,000,[2] and unsurprisingly, huge crowds flocked to see Cook's gibbet. For three days there was no abatement in public fascination with the spectacle, which also provoked outcry at the barbarity of the punishment. Fearing disorder, and sensitive to the disruption to travel and trade that the gibbet was causing the city, Cook's body was brought down by an executive order from the Home Office, and was buried, still in its cage, at the place where the gibbet had stood.[3] In Cook's case we see a rare—and possibly unique—instance of a formal reprieve being granted for the post-mortem element of the punishment for murder, though arguably the three days exposure was enough to satisfy the hanging in chains requirement. As a result, Cook's intact body both required and was granted burial. The body of William Jobling, whom we met in the last chapter, was also buried (mostly) intact in 1832, but not because of a formal reprieve. A guard had been set around Jobling's gibbet because the authorities were aware of how unpopular his punishment was with the local miners, and that a rescue attempt was a strong possibility. As soon as the guard had been lifted, three weeks after the gibbet was erected, Jobling's body was removed under cover of night. An account of the 'rescue' notes that it took so long to cut the cage down that there was no time to bury the body before daybreak.[4] The corpse, still secure in its iron cage, was retrieved from the place it had been hidden the next night and buried in an as-yet-unknown location.[5] In this case, Jobling's opportunity to rest in peace came at the hands of his friends and peers who acted against the law but in accordance with their own moral code. Jobling's corpse became a highly contested object: through his execution and gibbeting his corpse was thrust into a specific role in the maintenance of the power of the state and of those who controlled (and benefited most from) local economies. The destruction of the gibbet and burial of the body imbued Jobling's corpse with a certain power as symbolically important to a broader context of working class resistance.

The burial of a relatively intact body that had been gibbeted was a rare occurrence. Much more frequently, within the secure hold of the iron gibbet cage, as flesh began to putrefy and shrink, parts of the body would begin to drop to the ground. First the extremities—fingers, hands, feet—then parts of the legs and arms would fall and might be carried off by animals. Larger pieces became wedged in the cage such as the pelvis or ribcage and stuck there until more advanced decomposition caused connective tissues to wither. Then these, too, dropped to the ground. Usually the last part of the body left in the gibbet was the skull. Some of these pieces were taken away and put to other uses, as we will see below. There are also stories of family members of the gibbeted man visiting the site to collect the bones as they dropped. The mother of James Rook, executed in Sussex for robbing the mail in 1793, is said to have made repeated visits to his gibbet to gather her son's bones as they fell, and to take them (in secret) to a churchyard for burial.[6]

A small number of individuals sentenced under the Murder Act to anatomisation and dissection were not extensively cut or were spared this post-mortem punishment altogether. In these cases, intact and mostly intact bodies were disposed of by family and friends. The corpse of Earl Ferrers (the only member of the nobility convicted of murder during the life of the Act) was 'delivered to his friends for interment'[7] having been cut open and displayed to the public but not dissected any further. The corpse of 19-year-old Thomas Gordon was returned to his father. 'The surgeons', the newspapers reported, 'with great humanity gave up the body to the old man' and although Gordon had been sentenced to anatomisation and dissection, the body was buried intact.[8] At Surgeon's Hall, London, what remained after a corpse had been 'dissected to the extremities'—if anything remained—was buried in recycled coffin shells.[9] The remains of other corpses used for anatomical purposes were generally 'sewn together with a large surgical needle then wrapped in a woollen shroud used as a winding sheet and buried in a common grave [unmarked], normally no less than six deep'.[10] As Sarah Tarlow has found, there is archaeological evidence that attempts were made to give human remains from post-mortem investigations (i.e. autopsies) the semblance of a whole body before burial, as was the rule at the Edinburgh Royal Infirmary.[11] This was accomplished in a variety of ways, such as in the body found buried at St Peter's church, Barton-upon-Humber, in which the spine, organs, and ribs of an early nineteenth-century corpse had been removed, effectively leaving a 'skin bag'. Thanks to the insertion of a wooden stake in place of the spine, packing possibly with grass or moss, and sewing the remains together, a more human-looking 'body' was created for burial.[12] But many of the remains of criminal corpses created and punished under the Act were not laid to rest in these ways. Even in the cases in which the remains of Murder Act corpses may have been buried, burial was perfunctory and did not include family or friends of the deceased and probably paid only minimal attention to religious rites. A much more common outcome for those subjected to dissection under the Murder Act was to have as much as a third of a body's material ending up washed down the drains of the anatomy laboratories.[13] Other parts came to rest 'in pieces' in other forms and spaces.

PUT TO NEW USE: ANONYMOUS OBJECT OR UNIVERSAL REPRESENTATION?

The remains of those punished under the Murder Act were made to serve new, useful, and inventive purposes. These purposes sometimes relied on the remains having belonged to particularly notorious, infamous, or legendary people, but this was not always the case. Sometimes the bodily remains were preserved while the identity of the convicted was erased; often these criminal corpses did duty as representatives of a universal human body more generally. In such cases, the criminality of these bodies mattered only because it was this status, and more specifically the way the Murder Act left open the possibility for further uses that made them legally and ethically available for such ends. The fact that bodies produced under the Murder Act were meant to be the most heinous criminals, justifying the ongoing control of their bodies by the state (or medical men as agents of the state) meant not only that these bodies were infrequently buried, but also that they could be preserved without legal challenge or moral quandary. This was certainly the case in instances where Murder Act bodies were turned into educational objects that endured or were preserved far beyond the natural timeline of decay.

In the eighteenth century, anatomical dissection was a race against time as, in the absence of preservation technologies, bodies could decay past the point of utility in as little as three days. This urgency drove the tempo and determined the logistics of dissection for the penal surgeons who received bodies under the Murder Act, and decomposition often determined what parts of the body were used and over what time. The making of articulated skeletons for teaching and demonstrating purposes was a practical way to preserve and make use of the criminal corpse for a much longer period, long after the soft tissues had lost their utility.

As Anita Guerrini has argued, the spaces used by anatomists in early eighteenth-century London contained a minimum of one human skeleton.[14] These objects, and the scientific, religious, and symbolic meanings they carried, were an important part of creating an atmosphere of authority and authenticity for both practitioners and students of anatomy in this period. After anatomisation, public display, and any rapid dissection, bodies were reduced to bone by boiling, and then specialists connected and mounted the skeleton using wire. Unsurprisingly, given the key role the Murder Act played in supplying bodies for medical research, some articulated skeletons made from the bodies of people convicted and punished under the Murder Act were displayed in medical teaching spaces including Surgeon's Hall in London.[15] In addition to wired skeletons that could be used for instructional purposes, some surgeons preserved soft tissue samples from Murder Act corpses, using wax to maintain the shape and appearance of flesh or to highlight particular parts or processes. In these cases, it mattered little whose body part was preserved. As anatomical objects, criminal corpse parts were valuable for their ability to serve teaching and research purposes as representations of the human body, and the erasure of identity was a necessary part of the process that turned a convict into an educational object or artefact.

One criminal corpse created under the Murder Act was used for a curious project.[16] On 2 November 1801, Chelsea Pensioner James Legg was hanged at Newgate for the murder of fellow Chelsea Pensioner William Lamb.[17] In the month between the murder and the execution, three members of the Royal Academy of Arts, sculptor Thomas Banks and painters Benjamin West and Richard Cosway, made arrangements with the Chelsea Hospital surgeon Joseph Carpue—to whom the corpse was scheduled to be sent as Legg had been sentenced by the presiding judge to execution then anatomisation and dissection—to get possession of Legg's body when it was brought down from the gallows. But why did three artists want this criminal corpse so badly? The answer lies in longstanding debates over the physiological viability of Christ's crucifixion in the way it was typically portrayed by artists (Fig. 7.1). In question was whether or not the usual portrayal of the crucifixion involving nails being driven through the centre of the palms to attach the body to the arms of the cross was accurate. What Banks, West, and Cosway wanted to test was if it was possible for a body to be suspended in that way or if, as some argued, the weight of an adult male body would tear through the flesh. The alternative method for the crucifixion to have 'worked'

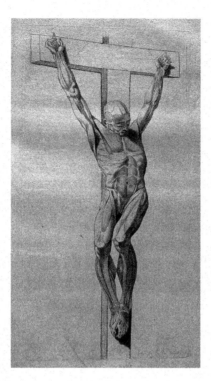

Fig. 7.1   Anatomically accurate drawing of an écorché (flayed) figure, by Jacques Gamelin, 1779 (Wellcome Collection)

was for the nails to be driven through the heel of the hand or the wrist where not only skin but bones and tendons would have allowed for successful suspension. Further, representations of Christ on the cross often involved meticulous physical detail with each muscle and sinew depicted to express the strain and agony of the torture. But how did this strain actually impact on the body, and with what visible result? These may seem like tiny details, but the ability to faithfully create visual representations of one of the most important moments for the Christian faith was certainly no small matter either for art or for religion.[18]

Legg's body was taken still warm from the gallows and hung by Carpue and Banks on a cross. Once it had settled into position and cooled, Banks made a cast of the whole body. Then, it was removed to

Carpue's anatomy rooms where he flayed the corpse, removing all the skin and exposing the body's muscles and tendons with the corpse still in the crucifixion position. Banks made a second cast to preserve the illustrative power of the test as, in the absence of sophisticated preservation techniques of later periods, the rate of decay that would render the example useless and the body dangerous was rapid. The casts were used and exhibited in the century that followed in medical and artistic spaces. By 1917, one cast—that of the flayed body on the cross—had been returned to the Royal Academy and it still hangs there today in the life-drawing room. There, it joins a collection of other anatomical casts, articulated skeletons, and anatomical drawings, and is used by members and students to improve their understandings and accurate representations of human physiology. In 2012, Legg's *écorché* was put on highly public display at the Doctors, Dissection and Resurrection Men exhibition at the Museum of London, demonstrating the enduring power of this corpse-based object to fascinate and educate the British public.

There was another practical purpose to which the remains of Murder Act corpses were put that did not depend on the identity of the murderer, but drew power specifically from their status as pieces of an executed body. Owen Davies and Francesca Matteoni investigated the use of criminal corpses as part of a healing tradition in eighteenth- and nineteenth-century England. During this time, it was popularly believed that the touch or stroke of a hanged man's hand (always male) had the power to cure skin disorders.[19] While the corpse still hung on the gallows, its hand was stroked three, seven, or nine times over the affected area of the individual who stood or was held up, in the case of small children, so the action could be performed. The hanged man's hand was made powerful through a combination of factors. Performing a selfless act at the moment of death might function as a sort of atonement for the sinner, drawing puissance from the religious associations of redemptive action. Or, according to ideas of animal magnetism (which also drove experiments in galvanism, as discussed in Chapter 5), the touch of the hanged man's hand held energy that could influence the flow of blood in the sufferer and help heal their complaint.[20] This 'cure' made use of a specific part of the corpses of those executed for capital crimes in Britain, including murder, but as it was only deemed effective while the corpse still hung on the gallows, it did not involve the preservation of parts of the criminal corpse. Other parts of the bodies of those executed by the state were, however, used for other medical purposes such as the use of the fat of executed criminals in Germany, Italy, France, and Spain for the treatment of injuries and wounds.[21] However, Davies and Matteoni have found no evidence that the fat from criminal corpses was collected in the dissection room in Britain, as had happened in France.[22]

The use of criminal corpses in folk medicine did not depend on the personal identity of the body. Their status as criminal corpses created under the Act made them available when other bodies were harder to get hold of during the eight decades that the Murder Act was in force. The notoriety of these individuals, and the details of their crimes and lives, however, did not create greater desirability of these objects or imbue them with greater value. But this was not the case for the parts to which we now turn.

Identity Matters/Identified Matter

Although there is some overlap with parts preserved for the sake of scientific interest, most of the human remains of Murder Act corpses that still exist today take the form of curios or artefacts whose value derives from their close connection with notorious and specific criminals. In eighteenth- and nineteenth-century Britain, some murderers achieved celebrity status, aided by the moral panic cultivated by print culture and periodicals. Sensational crime was the subject of extensive reporting and helped to sell both local and national newspapers. The *Ordinary of Newgate's Account*, a sister publication of the *Old Bailey Proceedings*, was published regularly from the late seventeenth to the late eighteenth centuries, and contained the biographies and last dying speeches of criminals executed at Tyburn in London. It was eagerly consumed by readers across Britain, as were broadsides and cheap pamphlets that relayed the sensational stories of gruesome and grisly murders and the retribution or justice meted out against those who committed such heinous (but fascinating) crimes. Just as the execution crowd clamoured to participate in the spectacle of execution and later in post-mortem punishments at the foot of the gibbet or around the table on which lay the anatomised criminal body, so too did people seek to connect more directly with relics of the condemned.

Body parts of murderers who had been punished under the Murder Act were sought after and turned into objects of curiosity, desire and fascination. But it was not their proximity to celebrity alone that made criminal body parts desirable. They were also visceral and dangerous things made safeish because their production was predicated on the death of the malefactor. Nevertheless, they still carried a sense of menace, and an allure based in part on revulsion but also an attraction to the extreme and unnatural nature of the individual capable of committing such heinous crimes. There was, at some level, a persistent belief that the 'dead body of the criminal retained something of the living individual's force and character'.[23] Thus possession of pieces of the criminal corpse allowed their owners to claim an exciting and titillating vicarious connection to the most dreaded of scoundrels.

Sarah Tarlow has traced the 'curious afterlives' of body parts of three individuals punished under the Murder Act.[24] Celebrity criminal Eugene Aram's skull was removed from his gibbet and became an object of great interest particularly as a phrenological test case. The skin and other body parts of notorious murderer William Burke (who with his accomplice William Hare killed at least 16 people in Edinburgh to sell the corpses to medical men) were preserved and coveted (Fig. 7.2). Finally, pieces of the corpse of William Corder, the infamous Red Barn Murderer, were

Fig. 7.2 Bust of William Burke, and pocketbook allegedly made from his tanned skin (Sarah Tarlow)

put on display in Scotland and England, and a book about the murder was bound in Corder's own skin. These body parts—or more properly these partial, selected, and preserved 'curios'—were the subject of attraction and financial exchange. The celebrity or notoriety of the individual from whom they were made conveyed a sort of glamour to those who possessed such items, or who saw or handled them.[25]

Because of the high value placed on them and their enduring appeal, artefacts and items created from Murder Act corpses are still on display in Britain, as they were in the eighteenth and nineteenth centuries. An encounter with one such object gave us the opportunity to reflect on historical processes of creation and preservation, and also the intensely personal experience of 'meeting' these human remains today. This object—the skull of John Bellingham—sits in a display case at eye level on the far right end of the main floor of Barts Pathology Museum, London. The skull is clearly special, and those with a knowledge of British political history will likely know why: on 11 May 1812 in the lobby of the House of Commons, Bellingham shot then Prime Minister Spencer Perceval in the chest. Having made no attempt at escape, Bellingham was immediately apprehended. Perceval died not long after. Bellingham was tried for murder on Friday 15 May 1812, convicted, and sentenced to death to be followed by anatomisation and dissection. This was, and remains, the only 'successful' assassination of a British Prime Minister. In accordance with the terms of the Murder Act, Bellingham was executed on Monday 18 May 1812 and his corpse sent to St Bartholomew's hospital where his post mortem punishment was accomplished.[26]

Today, this museum's extraordinary collection is housed in one immense room with three levels of shelves and walkways topped by a glass roof allowing light to all levels, and it is a key repository of anatomical artefacts. It serves both the curious public, to whom historical specimens and exhibits whose human material predates the restrictions of the Human Tissue Act (2004) are accessible on the ground floor, and medical professionals, who are able to access the two top floors where are kept items that can only be accessed by those currently engaging in medical education and research. Bellingham's skull sits in the area accessible to the public, presented alongside a reproduction of an image depicting the crime for which he was hanged, anatomised and dissected in 1812. The context is striking. The notorious nature of this artefact and the indelible link with criminality is on display for all to see. This skull is no anonymous didactic object of anatomical pedantry, valuable for its supposed unity with mankind or its metonymic ability to stand in for any human body. It is important because of the person of which it was once an integral part and the criminal act that brought the skull as artefact into existence. And the skull is not the only such object.

Most writings on Bellingham's story end at the gallows, but Carla Valentine and Helen MacDonald have begun to trace something of the fate of Bellingham's criminal corpse.[27] The notes of surgeon William Clift during the dissection of Bellingham's corpse found:

- The stomach contained a small quantity of fluid ('which seemed to be wine')
- The bladder was empty and contracted
- The penis 'seemed to be in a state of semi erection'
- The brain was found to be 'firm and sound throughout'.

Further, the surgeons were gratified to study the movement of the right auricle of Bellingham's heart for four hours following his execution, and then another hour when touched with a scalpel. A craniotomy was performed, hence the cut we still see today around the skull that allowed the top to be lifted off like a bowl. Indeed, Bellingham's skull (and not its contents) was the subject of some interest to phrenologists. This pseudoscience sought to identify physical signs of inherent criminality, and the drive to understand the workings of the individual who had committed such a violent and extraordinary crime was strong. Phrenologist George Combe wrote, 'The organ [of destructiveness] is large in the heads of cool and deliberate murderers. It is very large, and [the organ of] Benevolence small, in the skull of Bellingham, who murdered Mr. Percival. The temporal bones protrude at least half an inch in the situation of the organ of Destructiveness'.[28]

When the medical men of the Court had finished with Bellingham's body, the President, Sir William Blizard, gave it to a pupil of St Bartholomew's Hospital, a Mr. Stanley, and the body was broken up and distributed. MacDonald also notes that both the stomach and left testicle were sent to the College museum. Whether these specimens were preserved and if they still exist is unknown, but the skull—stamped with the number 159 and his surname at the front centre of the forehead—was. While phrenology is no longer considered a credible scientific field, it was far more accepted in the nineteenth century, and it was Bellingham's status as a notorious killer that brought his skull to the particular attention of phrenologists—rather than this skull standing in for the ubiquitous human skull, it instead served as an example of the 'organ of destructiveness' assumed to reside inside the worst criminals. Two hundred years later, there is little sign of accidental damage, and the object continues to draw public interest.

The context of the display of this skull as well as the history of its preservation mark it out as unusual or exceptional. Had Bellingham not murdered the Prime Minister, his body parts would most likely not have attracted specific interest or merited investigation or preservation. The fact of his crime made the skull valuable and intriguing. Encountering it unexpectedly in the present day called up feelings of fascination and revulsion that went beyond what might be expected for a more mundane skull. Joining the thousands of people who have experienced the thrill of proximity to Bellingham's skull over the past two centuries was a remarkable moment. It provided a visceral and material connection to a history we had read in accounts of the murder and its aftermath, and in that we are perhaps no different from the many others who have trodden the same path. While we stopped short of coveting the skull and desiring its ownership, the experience of spending time with this artefact gave us affective insight into why others might. At once grisly and harmless, alluring and revolting, the skull is inseparable from the authentic connection between the object and the commission of one of the most high-profile crimes of the early nineteenth century. Its possession might convey to the owner some of that fascination and power, and the satisfaction of mastery over the object, the individual, and their history.

Over the course of our research, the opportunity has not arisen to engage directly with a set of other, quite frankly, extremely disturbing objects created from the remains of Murder Act criminal corpses. As unnerving as it was to encounter Bellingham's skull stripped of flesh, we would have been more rattled to have encountered that flesh, preserved separately from the bones more usually located beneath. The skin of murderers was considered important and fascinating, and was both valued as a souvenir or curiosity, and also as a resource. The book covered in the skin of the Red Barn Murderer, William Corder, was not actually unique. Mary Bateman, a con-artist from Leeds, was convicted of murder by poison and sentenced to execution and anatomisation. In addition to the tip of Bateman's tongue being removed and added to the private collection of the governor of Ripon prison, two books were bound in her skin—both classic works from the sixteenth and seventeenth centuries.

There was no clear relation between the books and Bateman, and it is unclear why these works were chosen, but Davies and Matteoni have surmised that Bateman's 'reputation as a cunning-woman was key to why her skin accrued such a reputation'.[29] Corder and Bateman's skins were not the only ones to eventually be made into coverings for books, and in most cases the specific connection between the book and the murder is more explicit. As experts at Harvard working on the phenomena of books bound in human skin have recently stated, during our period of interest, 'the confessions of criminals were occasionally bound in the skin of the convicted'.[30] As Lindsay Fitzharris has noted, these items became 'objects of curiosity for the morbidly inclined'.[31]

Clearly, the skin of these executed convicts was seen to be important, and significant in a different way than simple access to flesh and bone for medical research and teaching. Skin is the part of the person most involved in social relationships: it is what we see, touch, and come to tightly associate with the people around us. The removal, preservation, and transformation of the skin of murderers like Corder and Bateman, among others, reads as an act of enduring torment and humiliation, consistent with acts of gibbeting and penal dissection. Unlike a skeleton, rendered anonymous by being stripped down and put to purposes independent of the specific actions and identity of the individual, books bound in the skin of convicts were—and are—notoriously and indelibly connected to the identity of the originator. Skin contains and identifies a body, as does the cover or binding of a book. Replacing a body's interior with text makes the book stand for the person—their body replaced by their story, but contained in the same skin. Covering a book in skin literally inscribes a story or narrative onto the body of the condemned, perhaps fixing in the minds of the public that no matter what friend, relatives or supporters might think, it is the story of murder and conviction before the courts that defined these individual lives in the grisly final calculus.

Mementoes such as these almost always came into circulation through the system of medical men. There are few, if any, known cases of people 'raiding gibbets for corpse pieces' or trying to make off with body parts of Murder Act criminal bodies during public anatomisation and dissection in Britain.[32] As discussed earlier in this chapter, it seems some friends and relatives patiently collected the bones of decayed corpses as they dropped out of gibbet cages, as Eugene Aram's wife allegedly did for the remains of her estranged husband's decaying body,[33] but we have found little evidence of people attempting intentionally to remove such pieces from the gibbet for personal gain. The structure of gibbets certainly played a role here, as they were designed to prevent interference with the corpse. Intrusion was discouraged using practices such as covering the very tall gibbet pole with spikes and tar, and with laws against interfering with gibbetted corpses that carried harsh punishments. One of the only incidences of this that we uncovered may well be apocryphal, but bears repeating. It is said that some young men removed the finger bones from the corpse of Spence Broughton as it swung on the gibbet.[34] These bones, it is alleged, were then ground into powder and used to make pottery in a nearby factory. Here again, the association with Broughton (or the supposed association as it was impossible to prove that this event actually happened) made these products valuable, intriguing, or desirable.

Enduring Power and Uncomfortable Questions

As Sarah Tarlow has noted, criminal bodies are powerful in a way that endures after death.[35] But not all criminal bodies are 'created' equal—the sentence of either gibbetting or dissection, for example, had a significant impact on how the public, collectors, and those searching for magical intervention were able to interact with the corpses. Bodies sent to the medical men were most often physically reduced until their utility for research and teaching were exhausted, and the remains disposed of with little of the ceremony or protocols usually afforded intact human bodies. For the parts that were preserved for purposes independent of identity, their power and utility inhered in their ubiquity. The persistence or survival of body parts of criminal corpses up to the present day, and the effort that has gone into their preservation, gives rise to opportunities to encounter direct physical remains of those punished under the Murder Act. Seeing first-hand or touching these objects, their physicality and authenticity (in contrast, perhaps, to the less tangible narrative remains that are the focus of Chapter 8) create a sense of making a direct and personal connection with these complex histories. Their continued existence also raises challenging ethical questions.

Ali Wells is curator of Natural Sciences and Human History for the Herbert Gallery in Coventry, situated in the West Midlands region of the United Kingdom. It is a city made famous in the twentieth century by the devastation wrought by targeted bombing in the Second World War, and the construction of a new cathedral next to the bombed-out ruins of the old, intended to serve as a symbol of peace and human unity in the post-war era. It is also home to a unique object: the head of murderer Mary Ann Higgens. In 1970 the Herbert Gallery acquired the head. In 2009 it was put on display there for the first time since its acquisition as part of The Hour of Death, an exhibition curated by Wells that examined the histories of the last two women to be hanged in Coventry (of whom Higgens was the penultimate).

We follow Wells in referring to Higgens's head and not her skull, as skin and cartilage remain, as does a waxy substance that was injected into the veins around her scalp. The presence of soft tissue, and not just bone, is what in part humanises this object and provokes questions about how it should be treated and displayed. We know from newspapers and the *Newgate Calendar* that although Higgens confessed to poisoning her uncle with arsenic, she had been compelled to do so by apprentice Edward Clarke who had extorted money from her and assaulted her whenever she failed to produce what he wanted. This context does not change the fact that Higgens was tried, found guilty, hanged on 11 August 1831, and afterward her corpse sent for anatomisation and dissection.[36] But it has impacted considerations of how her head should be displayed and whether human remains can or should be retained and treated as museum objects.

In legal terms, there were no impediments to the Herbert Gallery putting Higgens's head on public display as part of an exhibition. There were, however, ethical concerns. These were directed towards Higgens herself, a woman who was a victim in addition to being the perpetrator of a terrible crime. We can assume her remains have been retained and put to other uses without her consent, and as we will see in Chapter 9, such actions can visit harm on a person even after death. Wells was also concerned with those who might see the head and *how* visitors would see it. The context of display would influence whether audiences encountered the head as a fetishised and macabre object, an artifact of scientific interest, or as providing a visceral connection to a local and human life from the past and the complex historical context to which it relates. Probably all three of these possibilities would have a bearing on any encounter with the artefact.

The approach developed at the Herbert to the display of Higgens's remains takes into account concerns for both the living and the dead. Any display of her head is set within wider discussions of nineteenth-century crime and punishment (poisonings and crime detection in particular) and the history of medical access to human remains. Contextualised in this way, Higgens's head takes on a teaching role in addition to acting as an authentic and physical anchor for multifaceted histories that provoke reflection on the past but also the present. In The Hour of Death, visitors to the exhibition first encountered her story embedded in its historical context and were then able to choose whether or not to view her head, allowing individuals to decide how far they wanted to participate in the history and afterlife of these human remains. That the physical and narrative flow of the exhibition gave the option of seeing or not seeing the head created a moment for each visitor in which they had to make a decision, and therefore likely reflected—even if only briefly—about their relationship to Mary Ann Higgens's story, and the issues the existence of her preserved head raises in the present.

The criminal corpses produced and punished under the Murder Act possessed both utility and notoriety whether resting in peace or resting in pieces. As the curios and artefacts made from these criminal corpses that have been preserved up to the present and those that are still on display demonstrate, the criminal corpse remains with us, part of our entertainment or our education, as it has done for centuries. Whether object of curiosity, education, fetish, consumption or display, historical criminal corpses remain present and powerful. And as we will see in the next chapter, so too do their stories.

NOTES

1. In the case of criminal corpses sent for dissection, this contrasts with post-1832 practices when the bodies of the poor and unclaimed were made available for medical research and teaching under the Anatomy Act. As Elizabeth Hurren has noted, there were burial practices for these human remains, though they were not always followed/adhered to. For more See, Hurren, E.T. (2012), *Dying for Victorian Medicine: English Anatomy and Its Trade in the Dead Poor, c.1834–1929* (Basingstoke: Palgrave Macmillan).
2. Beazley, B. (2012), *Leicester Murders* (Stroud: History Press).
3. *The Newgate Calendar*, James Cook, Executed 10 August 1832, for the Murder of Mr. Paas, whose Remains he attempted to destroy by Fire.
4. For details See, Pelham, C. (1841), *The Chronicles of Crime, or the New Newgate Calendar* (London: Printed for Thomas Tegg).

5. Unsuccessful attempts have been made to locate Jobling's resting place, in the 1970s using a mine detector, and later by Vincent Rea, local historian and former curator of the Bede gallery at Jarrow, as stated in Tarlow, S. (2014), 'Technology of the Gibbet', *International Journal of Historical Archaeology*, Vol. 18, Issue 4, 668–699, 692.

6. It is believed that this story was the inspiration for Alfred Lord Tennyson's poem 'Rizpah' (1880). See for example, Simpson, J. (1973), *Folklore of Sussex* (B.T. Batsford, reprinted 2013).

7. *The Newgate Calendar*, LAURENCE, EARL FERRERS, Executed at Tyburn, 5 May 1760, for the Murder of his Steward, after a Trial before his Peers.

8. See, Gray, D. and King, P. (2013), 'The Killing of Constable Linnell: The Impact of Xenophobia and of Elite Connections in Eighteenth-Century Justice', *Family & Community History*, Vol. 16, Issue 1, 3–31.

9. See, Hurren, E.T. (2016), *Dissecting the Criminal Corpse: Staging Post-execution Punishment in Early Modern England* (London: Palgrave Macmillan), p. 139.

10. Ibid., quote at p. 153.

11. Tarlow, S. (2011), *Ritual, Belief and the Dead in Early Modern Britain and Ireland* (Cambridge: Cambridge University Press), pp. 91–92.

12. Rodwell, W. and Rodwell, K. (1982), 'St. Peter's Church, Barton-upon-Humber: Excavation and Structural Study, 1978–1981', *Antiquaries Journal*, Vol. 62, Issue 2, 283–315, 306. Quoted in Tarlow, S. (2011), *Ritual, Belief and the Dead in Early Modern Britain and Ireland* (Cambridge: Cambridge University Press), p. 92.

13. See, Hurren, E.T. (2016), *Dissecting the Criminal Corpse: Staging Post-execution Punishment in Early Modern England* (London: Palgrave Macmillan).

14. See, Guerrini, A. (2004), 'Anatomists and Entrepreneurs in Early Eighteenth-Century London', *Journal of the History of Medicine and Allied Sciences*, Vol. 59, Issue 2, 121–154.

15. See, Hurren, E.T. (2016), *Dissecting the Criminal Corpse: Staging Post-execution Punishment in Early Modern England* (London: Palgrave Macmillan).

16. An account of the case is described by the Royal Academy in an article about the cast taken from Legg's corpse, chosen as the 'object of the month', November 2012, available at http://www.racollection.org.uk/ixbin/indexplus?record=ART13544 (Accessed 25 July 2017).

17. Old Bailey Online, JAMES LEGG, Killing>murder, 28 October 1801. Available at https://www.oldbaileyonline.org/browse.jsp?div=t18011028-39 (Accessed 25 July 2017).

18. The artistic reputation of West had been knocked by the 'Venetian scandal' whereby he led a group of Academicians to be defrauded into using techniques wrongly touted as being used by Titian (the Plovis' formulae). His subsequent work was widely scathed by critics following his exhibition of two paintings displayed at the 1797 Academy Show, one of which he had used as inspiration for his St Georges Chapel 'crucifixion' stained glass window design. Two later drawings by West appear to demonstrate an 'atonement', in which he rejects the Plovis' formulae and instead adjusts proportions to match Legg's cast, with Jesus' and the thieves' shoulders abducted to 135°, and the digits flexed as they are in Legg's cast. For more on West's role in the experiment. See, Freshwater, M.F. (2015), 'Joseph Carpue's File Drawer Experiment—A Murder Mystery from 1801', *JPRAS Open*, Vol. 6, 74–85.

19. Davies, O. and Matteoni, F. (2015), '"A Virtue Beyond All Medicine": The Hanged Man's Hand, Gallows Tradition and Healing in Eighteenth- and Nineteenth-Century England', *Social History of Medicine*, Vol. 28, Issue 4, 686–705.

20. Ibid.

21. Davies, O. and Matteoni, F. (2017), *Executing Magic in the Modern Era: Criminal Bodies and the Gallows in Popular Medicine* (Palgrave Macmillan), p. 13.

22. Ibid., p. 17.

23. See, Tarlow, S. (2015), 'Curious Afterlives: The Enduring Appeal of the Criminal Corpse', *Mortality*, Vol. 21, Issue 3, 210–228.

24. Ibid.

25. Ibid.

26. See the essay by Carla Valentine, Curator at the Pathology Museum, Queen Mary University of London, 'The Skull of John Bellingham', available at http://www.qmul.ac.uk/pathologymuseum/bellingham/index.html (Accessed 25 July 2017).

27. See, MacDonald, H. (2005), *Human Remains: Dissection and Its Histories* (London: Yale University Press), pp. 17–19; Valentine, C. 'The Skull of John Bellingham', as above.

28. See, Combe, G. (1830), *A System of Phrenology* (London: Longman & Co.), pp. 176–177. In the same book, indeed on the same pages as the references to Bellingham, Combe mentions that the 'Organ of Destructiveness' is large in the skulls of several other murderers (who were convicted under the Murder Act), including Charles Hussey (executed 1818, Kent), James Nesbett/Nisbet (1820, Kent), George Lockey (1789, Yorkshire), Charles Rotherham (1817, Nottinghamshire), Robert Dean (1818, Surrey), and James Mitchell (1814, London).

29. Davies, O. and Matteoni, F. (2017), *Executing Magic in the Modern Era: Criminal Bodies and the Gallows in Popular Medicine* (Palgrave Macmillan), p. 9.
30. Quoted in Flood, A. (7 April 2014), 'Flesh-Crawling Page-Turners: The Books Bound in Human Skin', *The Guardian*.
31. Fitzharris, L. (31 January 2012), 'Books of Human Flesh: The History Behind Anthropodermic Bibliopegy', *The Chirugeon's Apprentice*.
32. Ibid., p. 7.
33. Tarlow, S. (2017), *The Golden and Ghoulish Age of the Gibbet in Britain* (London: Palgrave Macmillan).
34. Leader, R.E. (1901), *Sheffield in the Eighteenth Century* (Sheffield: Sheffield Independent Press), pp. 54–57.
35. See, Tarlow, S. (2015), 'Curious Afterlives: The Enduring Appeal of the Criminal Corpse', *Mortality*, Vol. 21, Issue 3, 210–228.
36. For details on Higgens's history, see Wells, A. (2016), 'A Hanged Woman and Her Journey to Becoming a Museum Object' [blogpost], *Harnessing the Power of the Criminal Corpse*, available at http://www.criminalcorpses. com/blog/2016/11/9/rest-in-pieces-the-story-of-a-hanged-woman-and-her-journey-to-becoming-a-museum-object (Accessed 25 July 2017).

CHAPTER 8

Folk Beliefs and Popular Tales

We suspect we are not the only people to misremember the plot of *Frankenstein*. The layering upon the novel of countless films and popular cultural references left us with a vague memory that the monster was created from criminal corpses, reanimated by the power of galvanism. Its subsequent pursuit of Frankenstein's friends and family was, we seemed to remember, the product of its atavistic, criminal nature bubbling through. It was a parable about the arrogance of a man trying to usurp the work of the creator, and the impotence of human design to reshape the essence of things and people.

On re-reading the novel, however, many years after we first encountered it, we realised how partial and distorted our memories of it were. The monster was created using chemistry and, it is implied, the already discredited alchemical ideas of Cornelius Agrippa and Paracelsus. Author Mary Shelley is unspecific about where the materials come from, beyond saying that Frankenstein 'collected bones from charnel-houses and disturbed, with profane fingers, the tremendous secrets of the human frame'.[1] The monster's viciousness was a result of hurt and anger, produced by his cruel exclusion from human communities of love, not an expression of a criminal essence, though Victor Frankenstein fails to understand this.

It was therefore no part of Shelley's vision that criminality inheres in the body. Nevertheless, that idea has been attractive to the authors of

© The Author(s) 2018   213
S. Tarlow and E. Battell Lowman, *Harnessing the Power of the Criminal Corpse*, Palgrave Historical Studies in the Criminal Corpse and its Afterlife, https://doi.org/10.1007/978-3-319-77908-9_8

later reworkings of the Frankenstein narrative. At issue is a key question about what constitutes a criminal corpse: the origin and nature of criminality. This was a divisive problem in the nineteenth century. It was essential to Enlightenment ideas of human character and behaviour that people are shaped by their environments. They are thus susceptible to education and capable of almost infinite improvement.[2] Robert Owen, for example, published a series of essays on human character, developing the principle that 'Any general character, from the best to the worst, from the most ignorant to the most enlightened, may be given to any community, even to the world at large, by the application of proper means'.[3] On the other side of the table, especially as the nineteenth century progressed, were the adherents of the developing sciences of human behaviour: the disciplines which would eventually become anthropology and psychology. Their precursors, phrenology, physiognomy and early biological anthropology relied on anthropometry to reveal capacity, ability and character. These were very much written in the body, inescapable and congenital. Criminality in this paradigm was an inevitable quality of 'bad blood', poor breeding that no amount of better housing and universal education would be able to eradicate.

The implications of taking an 'environmental' or 'congenital' view of criminality are great and this book is not the place to open them up much further. However, the currency of this debate in the nineteenth century affected and informed many of the cultural products of the period. The political sympathies of authors are manifested in their approach to criminality: environmentalists being inclined to stories that emphasise the difficult conditions which give rise to criminal behaviour; more conservative stories focusing instead on the early manifestations of individual Bad Character in people who matured into full-blown villains.[4]

The power of the executed corpse can be channelled not only through the deployment of the material body and its constituent parts, though that has been the focus of most of this book so far, but also through its evocation in stories, visual media and other forms of representation. This chapter looks at the cultural afterlives of criminal bodies. While some of these are the histories of body parts that have been reappropriated and transformed, such as the curious power of the hanged man's hand, considered in the previous chapter, other afterlives are the products of a creative cultural imagination: songs, stories, plays and films.

THE CRIMINAL CORPSE IN LITERATURE

During the period of the Murder Act the executed body in a gibbet seems to have caught the imagination of writers to a greater degree than that of the body under dissection. This imbalance was largely reversed in the periods before and after this. Early modern writers were obsessed with the process of dissection and anatomisation; anatomy was a ubiquitous metaphor in the poems, plays and essays of the period, as demonstrated by Jonathan Sawday,[5] and in the twentieth and twenty-first centuries the interest of cultural historians in opening bodies has informed the work of novelists and others too. Gibbets, on the other hand, have been almost absent from the historiography, and their representation in popular culture, when it occurs, is a mishmash of misunderstandings, a pastiche of a homogenous 'olden days' in which pirates are left to starve in birdcages along the coast.

Stories of criminal corpses and their fates occur throughout the post-medieval period. These range from the well-informed and profound ruminations of Donne in the early modern period, to anonymous ballads and pamphlets.[6] This chapter explores some of these literary afterlives.

The Gibbet in Literature

John Grindrod was gibbeted in 1759 on Pendleton Moor in Lancashire for poisoning his wife. The ballad 'Old Grindrod's ghost', collected by William Ainsworth in 1872, tells a story that recurs with variants around the country. The full text of the ballad is as follows:

> Old Grindrod was hang'd on a gibbet high,
> On a spot where the dark deed was done;
> 'Twas a desolate place, on the edge of a moor,
> A place for the timid to shun.

> Chains round his middle, and chains round his neck.
> And chains round his ankles were hung;
> And there in all weathers, in sunshine and rain,
> Old Grindrod the murderer swung.

> Old Grindrod had long been the banquet of crows,
> Who flock'd on his carcase to batten;
> And the unctuous morsels that fell from their feast,
> Served the rank weeds beneath him to fatten.

All that's now left of him is a skeleton grim.
The stoutest to strike with dismay;
So ghastly the sight, that no urchin, at night,
Who can help it, will pass by that way.

All such as had dared, had sadly been scared,
And soon 'twas the general talk,
That the wretch in his chains, each night took the pains.
To come down from the gibbet—and walk!

The story was told to a traveller bold.
At an inn near the moor, by the host;
He appeals to each guest, and its truth they attest,
But the traveller laughs at the ghost.

'Now to show you,' quoth he, 'how afraid I must be,
A rump and a dozen I'll lay.
That before it strikes one, I will go forth alone,
Old Grindrod a visit to pay.

'To the gibbet I'll go, and this I will do,
As sure as I stand in my shoes;
Some address I'll devise, and if Grinny replies,
My wager of course I shall lose.'

'Accepted the bet; but the night it is wet,'
Quoth the host. 'Never mind,' says the guest;
'From darkness and rain the adventure will gain
To my mind an additional zest.'

Now midnight had toll'd, and the traveller bold
Set out from the inn all alone;
'Twas a night black as ink, and our friend 'gan to think
That uncommonly cold it had grown.

But of nothing afraid, and by nothing delay'd,
Plunging onward through bog and through wood.
Wind and rain in his face, he ne'er slacken'd his pace,
Till under the gibbet he stood.

Though dark as could be, yet he thought he could see
The skeleton hanging on high;
The gibbet it creaked, and the rusty chains squeaked,
And a screech-owl flew solemnly by.

The heavy rain patter'd, the hollow bones clatter'd.
The traveller's teeth chatter'd—with cold—not with fright;
The wind it blew lustily, piercingly, gustily;
Certainly not an agreeable night!

'Ho! Grindrod, old fellow!' thus loudly did bellow
The traveller mellow,—'How are you, my blade?'
'I'm cold and I'm dreary; I'm wet and I'm weary;
But soon I'll be near ye!' the skeleton said.

The grisly bones rattled, and with the chains battled;
The gibbet appallingly shook;
On the ground something stirrd, but no more the man heard—
To his heels on the instant he took.

Over moorland he dash'd, and through quagmire he plash'd;
His pace never daring to slack;
Till the hostel he near'd, for greatly he fear'd,
Old Grindrod would leap on his back.

His wager he lost, and a trifle it cost;
But that which annoy'd him the most,
Was to find out too late, that certain as fate,
The landlord had acted the ghost.

The story of the boastful man at the inn getting his comeuppance at the foot of the gibbet is a widely known one, and variants relate to other gibbets and other vainglorious travellers. The same story is told of Matthew Cocklane, executed in Derby in 1776 and others.[7] A letter to *The Mirror of Literature, Amusement and Instruction* locates the story to Craven and the gibbet of 'Tom Lee', and the victim of the joke is a retired sergeant, given to rambling and boastful accounts of his military career.[8] The *Mirror*'s correspondent pertinently remarks that the murderer had almost as strong and frightening a hold over the imaginations of ordinary people after he had been gibbeted as he had before he was caught: 'Though the strong arm of the Law had incapacitated this desperado from any further molestation of person or property, yet over the minds of the superstitious and the ignorant, he seemed to have a greater dominion than ever'.[9] It is a popular piece of gibbet lore, perhaps especially in the nineteenth century when actual gibbetings were very infrequent and their occasions drew huge crowds and much public interest.

The ballad or broadside was the cheapest, most popular, and least accomplished literary form during the period preceding and during the Murder Act. Broadsides of true crime and punishment were perennially popular, even into the later nineteenth century (and indeed the True Crime sections of modern bookshops remain their large and well-frequented successors). In a tradition spanning from the sixteenth to the nineteenth centuries, cheap, sensational accounts of the crimes and fates of notorious offenders were produced rapidly to coincide with the peak of public interest in the case. Typically they contain a ballad of poor literary quality and, especially in the eighteenth and nineteenth centuries, a prose account of the affair. Illustrations, when they occur, are generic woodcuts; frequently the same prints are used repeatedly for decades or even centuries. Broadsides about mid-nineteenth-century criminals might be accompanied by a woodcut of individuals in seventeenth-century costume. Interestingly, with the end of hanging in chains and the gradual removal of punishment from public view, crime ballads also came to focus more on the crime and its forensic solution, while executions disappear from the story after around 1830.[10]

Literary treatments of real criminal cases were not uncommon throughout the period, though the dead body or post-mortem punishment of an executed criminal features less frequently. Eugene Aram, for example, who was hung in chains following his conviction for murder in 1759 became a celebrity more than sixty years after his death because of the publication of two massively popular literary works about his life and execution in the 1830s. Eugene Aram was an educated man, working in a school in Norfolk, when he was arrested for a murder that had taken place fifteen years earlier, and in a case where no body had been found. A former associate of Aram's named him as the culpable party and, on very slender and unreliable evidence, Aram was found guilty and sentenced to be hung in chains after execution. Fictionalised retellings of his life, crime, flight from justice, and eventual trial and execution were produced from the imagination of poet Thomas Hood and popular novelist Edward Bulwer Lytton. Thomas Hood's narrative poem 'The Dream of Eugene Aram' was first published in 1831 and is a ruminative account of Aram reflecting on his life and crime in the moments before his arrest, finishing with him being taken away for trial.[11] Bulwer Lytton's *Eugene Aram*, published the following year,[12] is ludicrously melodramatic and sentimental for modern tastes but nineteenth-century English speakers around the world lapped it up. These two stories of Eugene Aram were both bestsellers; the novel was adapted for the stage and was the inspiration for a series of prints by Gustav Doré. Aram thus remained the object of popular interest well into the twentieth century. PG Wodehouse even has Bertie Wooster quoting Hood's poem in proper Wooster style:

> All I can recall of the actual poetry is the bit that goes: Tum-tum, tum-tum, tum-tumty-tum, I slew him, tum-tum tum![13]

However, in neither of these works does the executed body of Aram play a significant role. Hood's poem finished before Aram's trial, and Bulwer-Lytton's novel leaves Eugene at the moment of his death. His sentencing is not mentioned, beyond execution, and neither Aram himself nor any of the other characters reflect on hanging in chains.

During the nineteenth century, even after hanging in chains had mostly died out as a punishment, the continuing presence of gibbets in the landscape seems to have made the gibbeted body a powerfully meaningful and memorable part of local landscapes, and as such it is frequently described to add atmosphere or to serve a symbolic role in nineteenth-century literature.

The body itself could be entirely gone or reduced to bones without the material structure of the gibbet losing its power to terrify and fascinate. The opening paragraphs of Charles Dickens's *Great Expectations* refer to young Pip's experience of the gibbet on the marshes where he encounters the escaped convict Magwich.[14] The gibbet is a landmark in the featureless wetlands, even though the name of the man who hung there is forgotten or unspoken.[15] It is emblematic, even anonymously, of the consequences of crime that haunt and drive the plot of the novel. By contrast, anatomised bodies barely feature in Dickens's work, though the procurement of (noncriminal) corpses for dissection was part of the livelihood of Gerry Cruncher in *A Tale of Two Cities*.[16]

For William Wordsworth in 'The Prelude', the gibbet appears as a significant childhood memory. An unexpected encounter with a gibbet during an adventurous expedition constitutes one of the poet's first shocking and macabre encounters with mortality:

> ... and I fled,
> Faltering and faint, and ignorant of the road[17]

Duncan Wu notes that 'Nicholson's gibbet [the one that best fits the geography of Wordsworth's childhood recollections at this point] had not 'mouldered down' in 1775, and a 5-year-old would not have ridden that far'.[18] Kelley considers that the gibbet was either at Gibbet Moss near Hawkshead or Penrith Beacon.[19] However, it is not the historical accuracy of this episode that matters as much as its symbolic power. The boy Wordsworth is both innocent and naïve; the gibbeted criminal represents the worst of human evil, and his punished corpse is a sign of the brutality of social and political institutions. The presence, in the midst of a beautiful, rural landscape, of a decaying body, and the body of a murderer at that, disrupts any naïve pastoral idealism, and gives a dangerous, Romantic force to the hills and lakes of the young poet's early life.

Perhaps the fullest nineteenth-century fictional account of a child's encounter with an English gibbet was not written by an English person at all, but a Frenchman, albeit in a novel set in England. Victor Hugo's *L'homme qui Rit* (literally, The Laughing Man, though published in English as '*By Order of The King*') was first published in 1869. The story follows the life of Gwynplaine. Stolen as a baby and mutilated by having the corners of his mouth cut up towards his ears to give him a permanent grin,[20] Gwynplaine finds himself wandering alone in southern England. Eight-year-old Gwynplaine's encounter with the gibbet takes place at dusk on a winter's day as the child walks along the cliffs of the south coast of England. Although Hugo's novel was set in England, the author himself was, of course, French, and his account of eighteenth-century practice, a century later and a country removed, might not be entirely accurate. However, references to Jack the Painter,[21] and some other historical details, such as the three men still hanging in chains at Dover castle in 1822,[22] suggest that this chapter was well researched.

The gibbeted man encountered by young Gwynplaine was tarred and partially decayed. However, Hugo describes traces of repair and maintenance that had been carried out on the body, suggesting that it had been freshly retarred. The tar here functions to prolong the existence of the man: 'They had not cared to keep him alive, but they cared to keep him dead.'[23] Hugo suggests that extending the man's tenancy of the gibbet would postpone the moment at which a new example must be made. The coastal gibbets of smugglers were intended, says the narrator, to act as 'beacons' to other smugglers, although these beacons did not deter crime, and Hugo's choice of word (réverbère, which as a noun means lamp-post or streetlight) implies instead that the conspicuity of the gibbet might have inadvertently assisted the smugglers in finding their way along dangerous coasts with few landmarks. The two most striking aspects of this encounter are, first, the way the gibbeted corpse occupies a liminal space between life and death; and second its multisensory apprehension by the onlooker. Gwynplaine's encounter with the gibbet is an immersive and bodily experience.

The body, as the child apprehends it, is neither wholly alive nor wholly dead, but instead exists in some terrifying liminal state. Though skeletal, and its eyes, nose and mouth no more than 'holes', his teeth 'retained a laugh' and his drooping head seemed somehow alert.[24] The child feels keenly an absolute horror of being regarded by blind eyes that 'seem to have vision', as he succumbs to a debilitating terror. Then the wind begins to blow the gibbeted man back and forwards; there is, observes Gwynplaine '[n]othing stranger than this dead man in movement.'[25] As the dead man swings faster and further, the chain from which he is suspended makes a disturbing grinding noise, like breathing. As the wind increases the grinding cry becomes a shriek. Then, out of the gathering evening and storm, a flock of ravens appears[26] and the birds perch first on the gibbet and then on the corpse itself. In a particularly grotesque passage, the black birds swarm over the body, which is now a writhing form of black wings, beginning to move with renewed energy in the returning storm. The body seems to struggle in 'convulsions' and Gwynplaine has the impression that the man is trying to escape his cage, 'possessed with hideous vitality.'[27]

Hugo's gibbeted man occupies an intermediate space between life and death. Though a dead thing, it moves and makes noises like a living person. Its blind eyes can see; its empty mouth laughs at him.

Looking back on the gibbet from the later nineteenth century, A. E. Housman is almost nostalgic for the gibbet which formerly stood 'Fast by the four cross ways'. He contrasts the place where a hundred years before the dead criminal 'stood on air' above the moonlit sheep, with the mechanical long drop hanging inside the walls of Shrewsbury Jail.[28] The gibbet of Housman's imagination fits into a romantic landscape of 'moonlit heath and lonesome bank' more naturally than the quick end of the prison execution, accompanied by the 'groan' and whistle of trains that run through the night. Although Housman adds a footnote to explain that 'keeping sheep by moonlight' was a euphemism for hanging in chains, it is not an expression we have encountered elsewhere.

A fictional gibbet appears in the children's moralising book of religious education *The History of the Fairchild Family*, written by Mrs. Sherwood and published in 1853. The nauseatingly pious Mr. Fairchild takes his children, aged seven, eight and nine, to see the remains of a murderer hanging on a gibbet a few miles from their home. This trip is intended to act as a cautionary tale, following a squabble between the children. Their father tells the story of how Roger, the man whose remains they have come to see, was brought up in a misguidedly permissive household and eventually quarrelled with and killed his own brother. Although the children have already been punished for their argument, and have expressed their regret, Mr. Fairchild decides they should be taken to the gibbet at Blackwood to see something 'they will remember as long as they live: and I hope they will take warning from it, and pray more earnestly for new hearts, that they may love each other with perfect and heavenly love'.[29]

The Melancholy of Anatomy

The literary weight and significance of anatomical dissection has received extensive critical attention, especially the cultural production of the early modern period. However, the criminality of the body that is being dissected is not necessarily, or even normally, considered in literature. Anatomical dissection was not only vastly interesting in itself to the writers and artists of early modernity, the process of dissection—opening, laying visible, describing and exploring in an ordered fashion—became a defining metaphor in the period. 'Anatomies' of subjects as diverse as Popish Tyranny, Wit and Fortune were published in the period 1576–1650. Richard Sugg lists 78 titles containing the word 'anatomy' in a metaphorical sense dating to that period, and a further 44 which use 'anatomise' as a verb.[30] As Sawday has described, this kind of anatomy is analogous to geographical exploration: a noble, courageous and selfless endeavour carried out to the betterment of humankind.[31] To anatomise is to know, describe and map in detail an area that had been improperly, incompletely and impressionistically known before. It is itself a 'civilising' act, in the sense of bringing the unknown realms of Nature or the Mind of God into the control of Man. The tension between knowledge that is properly God's and that which should be brought into human understanding is an important motor of early modern writing on the subject of anatomy.

There was nothing necessarily criminal about the dissected bodies of these anatomies, but a dread of being anatomised was harnessed to drive both moralising tracts and works of comedy. Both Sugg and Sawday mention popular comedies and melodramas, such as *The Atheist's Tragedy* (Cyril Tourneur 1611) and *The Anatomist* (Edward Ravenscroft 1696), which employ a threat of anatomisation to give urgency and an entertaining frisson to the action. In these popular works the would-be anatomiser has no nobility of purpose. Instead, the 'Atheist' of Tourneur's work is driven by a heretical desire to uncover knowledge that properly belongs to God alone. The central character requests the body of his courageous nephew so that he can find 'in his anatomy' where his bravery is located. This demonstrates both that a belief in the somatic basis of character could be articulated in 1611, and that such a belief was condemnable as impious and immoral.

In the early eighteenth century, the corpse under anatomical exploration could provide an element of dark, transgressive humour in popular culture, juxtaposing the solemn and forbidden nature of the corpse with a trivial personal or romantic goal. In Alexander Pope's satirical *Memoirs of the Extraordinary Life, Works, and Discoveries of Martinus Scriblerus*, published in 1741, pedantic scholar Scriblerus tries to acquire a criminal corpse in order to study anatomy. The servant tasked with its procurement is trying to drag the body secretly upstairs. As the body threatens to slip from his grasp, he tightens his grip around its abdomen. This has the unfortunate effect of causing the cadaver to expel a quantity of gas in a loud and vibrant fart. In terror, the man abandons his job and runs away, leaving the neighbours, aroused by his shrieks, to discover a corpse on the stairs. Assuming that a murder has been newly discovered, they summon the watch, who capture Scriblerus and his unfortunate servant. The would-be anatomists are taken to see the Justice, who asks what their profession is. The servant declares, in an abominably poor attempt to exonerate them, 'It is our business to imbrue our hands in blood; we cut off the heads, and pull out the hearts of those that never injur'd us; we rip up big-belly'd women, and tear children limb from limb'.[32]

Where dissection or anatomy is featured in nineteenth-century fiction, the emphasis is invariably on the dreadful experience of the cadaver, or the ghoulish appearance of the corpse as a thing to terrorise the living, to horrifying or comic effect. There is no sense here that the criminal bodies are being sacrificed to further some higher goal or greater good of improving medical care or developing surgery. The exploration of the body's interior in the time of the Murder Act is no longer the noble, humanistic voyage celebrated by Donne two hundred years earlier (Donne consistently uses the language of geographical exploration to describe knowledge of the body). It is desecration motivated by prurience.

The anatomised corpse rarely appears in fiction, except as an object of comedy or horror. In deSade's *La Marquise de Gange* (1813), for example, the eponymous marquise is being held prisoner in an old castle. She notices that the door to a previously locked room has been left ajar and, driven by curiosity, goes into the room. Inside she encounters a 'horribly mutilated' cadaver, which was in the process of being dissected in a private anatomy room.[33]

The Magical Corpse

The dead body has specific powers in early modern literature, particularly judicial and curative powers. The medicinal power of the body in history has been extensively explored in history by Sugg and by Davies and Matteoni, and in literature by Robert Brittain.[34] The judicial power of the dead body lay not so much in the criminal corpse as in the body of the victim. The belief that the cadaver of a murder victim would bleed afresh in the presence of its murderer was known in ancient texts,[35] but occurs in historically documented cases until the nineteenth century, and is important to the plots of Shakespeare's *Richard III*,[36] and Walter Scott's *Fair Maid of Perth*.[37] That it was well known beyond Britain is evident in Mark Twain's *The Adventures of Tom Sawyer*, when Tom hopes that the bleeding of Dr. Robinson's body would turn suspicion towards Injun Joe, as the dead man was lifted onto a wagon.[38] Davies and Matteoni point out that the judicial power of the phenomenon of post-mortem bleeding (properly called 'cruentation'), demonstrates a belief in the sympathetic link between the dead victim's body and that of the living criminal.[39]

The medicinal power of the criminal corpse, considered in the previous chapter, is less explored in literature, though the quest to obtain a touch of the dead man's hand provides the narrative drive in Thomas Hardy's short story *The Withered Arm*. In that story, a woman, sensing that her new husband is preoccupied and seems less interested in her, concludes that he feels revulsion at her withered arm and seeks out a cure. As orthodox medicine has failed her, she secretly goes to the assizes in the county town in order to apply the touch of the dead man's hand, and negotiates access to the body of a freshly executed criminal. Unexpectedly she meets her husband by the side of the cadaver: the hanged man is her husband's son from his first marriage, and keeping quiet about his knowledge of his son's fate, rather than distaste for his wife's deformity, explains his subdued demeanour.

The hand of the hanged man had other powers beyond the curative. The tradition of the 'hand of glory' is well known in folk stories (Fig. 8.1). Shane McCorristine gives a version of the tale as follows, though there are numerous variants known as ballads or collected by folklorists in the nineteenth century.[40] A traveller asks to leave a box overnight in the house of a wealthy farmer. In the middle of the night, the maidservant, unable to sleep, goes downstairs and is alarmed to see a tall man remove from the box a withered human hand. He then proceeds to fix the hand to a board, and smear it all over with some kind of grease, and then to set fire to the fingers so that they burn like candles in a candelabra. The man then begins to burgle the house. The maid runs upstairs to wake the sleeping members of the household but is unable to rouse

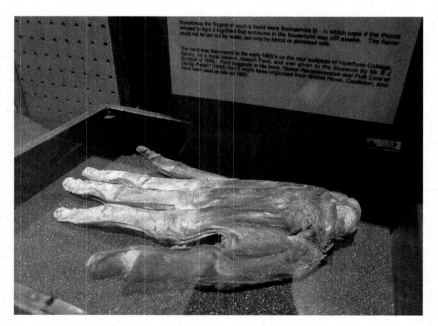

Fig. 8.1 Hand of Glory, Whitby Museum (Sarah Tarlow)

them at all. Returning downstairs, she tries to extinguish the burning hand, but water seems to have no effect. Remembering something she had once heard, she takes a jug of milk and throws that over the hand of glory which happily extinguishes the flames. Now the enchantment is over, the family wakes up and the thief flees, empty-handed.

Details differ from this version, which was collected by a correspondent writing about the folklore of Cheshire in 1872. In other versions, the burglar is a traveller disguised as a woman, seeking shelter at an inn, or the hand acts as a candle holder for a candle made from human fat and other magical ingredients, rather than burning itself; but the details of the nefarious use of the hand, its origin as part of a hanged criminal, the unwakeable sleep of the inhabitants and the resourcefulness of the servant remain the same. The power of the hand of glory was sufficiently well known that it featured in works of nineteenth-century literary fiction, including Walter Scott's *The Antiquary* (1816) and Richard Harris Barham's *The Ingoldsby Legends* (1840). Scott actually gives detailed instructions for how to make a hand of glory, which he puts into the mouth of Westphalian Mr. Dousterswivel:

> it is hand cut off from dead man as has been hanged for murther, and dried very nice in der shmoke of juniper wood, and if you put a little of what you call yew wid your juniper, it will not be any better—that is it will be no worse—then you do take something of de fatsh of de bear, and of de badger, and of de great eber, as you call de grand boar, and of de little sucking child as has not been christened (for dat is very essentials), and you do make a candle, and put it into de hand of glory, at de proper hour and minute, and with de proper ceremonish, and he who seeksh for treasuresh shall never find none at all.[41]

The power of the criminal corpse could, alternatively, be mediated by objects or things that had 'caught' the power of the body itself. The body's power could pass into other things through contagion, proximity or sympathy. The power of the hangman's rope has been discussed by Matteoni, and Davies and Matteoni.[42] Hangmen were able to make some money by selling lengths of the rope used for a hanging as a remedy or safeguard against illness. Matteoni and Davies also note the associations of the criminal corpse with the magical and medical power of the mandrake, which was important in some works of literature. The mandrake grew, by tradition, beneath the gallows, where it was nourished by, and imbued with power from, the corpse hanging above it and dripping blood or other powerful bodily fluids onto the ground. The mandrake tradition appears in European literature from around 1500.[43] In Ludwig Achin von Arnim's Gothic tale Isabella of Egypt, a beautiful Gypsy princess creates a magical creature—a mandragore—by bringing to life a mandrake that had grown beneath the gallows on which hung the body of her father, fed by his tears. The mandrake creature—called Mandragore—is an evil being but in the service of Isabella, and has the ability to find buried treasure, of which Isabella needs a plentiful supply if she is to achieve her goal of marriage to the crown prince.

During the nineteenth century, the dead criminal continued to exert power and to be co-opted into other ideological projects, especially within the growing spiritualist movement. Spiritualists believed that the dead occupied an imminent geography 'beyond the veil' or 'on the other side'. Communication between the dead and the living could be facilitated by mediums who were sensitive to the presence of dead souls among us. Among the spiritualists, the traditional opportunity for the about-to-be executed to express last-minute repentance of their evil deeds could now be extended into the post-mortem period. The executed could, through a medium, express remorse, and give moral or practical instructions to their friends and family. For example, William Saville was executed in 1844 in Nottingham for the murder of his wife and three children. Indirectly, he was responsible for the deaths of many more, as at least 16 people were crushed to death in the crowd that surged along Nottingham's narrow streets and down its stone steps following Saville's execution in front of Shire Hall. Saville's spirit, however, appeared to spiritualist medium John G.H. Brown in a crystal ball. He confessed to his crimes and described his existence in the 'lower regions' (a kind of 'Hell-lite'[44] for a sect that did not believe in everlasting damnation, but required a place where sins could be paid for and regretted).[45] Spiritualists generally opposed capital punishment altogether on the grounds that God had determined an allotted span for every person, and curtailing that period left the spirit of the departed in a kind of protestant limbo until the time of what would have been their natural death.

THE CRIMINAL CORPSE IN ART

Many artistic depictions of criminal bodies are incidental to other illustrative purposes: gibbets on maps and town plans, for example, or heads on spikes above a gate, bridge or wall in early modern town views.

In the Middle Ages, and indeed afterwards, the referent for almost all depictions of the criminal corpse was the body of the crucified Christ, or the bodies of the criminals who died beside him, as was discussed in Chapter 2. The cross casts a long shadow, and well into the early modern and modern age, depictions of criminal bodies were self-consciously positioned with reference to the very deep tradition of presenting the criminal body as at once abject and an object of salvation.

The anatomised criminal corpse has a lower profile in textual afterlives than in visual art. The changing depictions of anatomical dissection are evident in a comparison between two very well-known and much reproduced images, separated by 120 years: Rembrandt's painting *The Anatomy Lesson of Dr. Nicolaes Tulp* (1632) and Hogarth's etching *The Reward of Cruelty* (1751). That the first of these is a commissioned group portrait, worked in oils, and the second a popular grotesque is in itself significant. What had been possible to represent heroically in the seventeenth century was now quite the reverse.

In early modernity, some of the best-known and most widely reproduced images are Rembrandt's paintings of bodies under dissection. His *Anatomy Lesson of Dr. Nicolaes Tulp* is a commissioned group portrait, worked in oils. In this picture, the object of study is the body on the table, but the body here is sympathetically represented. While the body is clearly inanimate, and contrasted with the lively curiosity of the gentlemen gathered around, the prostrate and pale form, covered with a loin cloth, is reminiscent of Christ taken down from the cross. The drooping head and foreshortened perspective of the criminal body in Rembrandt's other well-known anatomy painting, *The Anatomy Lesson of Dr. Joan Deyman*, further evokes Christ on the cross. The sacrificial and redemptive associations of the criminal corpse are clear in both of these pictures, but so also is the heroism of the anatomist.[46] Heckscher argued that Rembrandt worked with the gentlemen commissioning the portrait of themselves with Dr. Tulp to represent the doctor as a transformative and benevolent figure, transforming the dangerous, criminal, deviant threat to ordered society into progressive and benevolent knowledge.[47]

Such a representation was no longer possible in the cynical age of Hogarth. 'The reward of cruelty' is the fourth of a series of four etchings depicting the life of fictional anti-hero Tom Nero (Fig. 8.2). His childhood cruelty to animals has matured into violent and murderous cruelty to people, and in this final episode an ironic reversal of fortune has occurred and Nero's own body has become the subject of violent action. The medical men who crowd around the dissection are taking

THE REWARD OF CRUELTY.

Behold the Villain's dire disgrace! | Torn from the Root, that wicked Tongue, | His Heart, expos'd to prying Eyes,
Not Death itself can end. | Which daily swore and curst! | To Pity has no Claim:
He finds no peaceful Burial-Place, | There Eyeballs from their Sockets wrung, | But, dreadful! from his Bones shall rise,
His breathless Corse, no friend. | That glow'd with lawless Lust! | His Monument of Shame.

Fig. 8.2 *The Reward of Cruelty.* William Hogarth 1795 (Wellcome Collection)

a greedy and unseemly joy in the destruction of Nero's body. Although dead, the body is being hoisted up so that he seems to be responding to the physical discomfort and indignity of the procedure. There is a comic and grotesque suggestion that the body is being prepared as food.

A cauldron of boiling bones is evidence that the remains of the current subject of study will join the articulated and mounted specimens displayed in niches around the room. Disgustingly, a dog is nosing at the man's heart and entrails: canine revenge for Nero's juvenile torture of an unfortunate puppy. Hogarth's dissection scene is a thousand miles from the respectable, scholarly anatomies of Rembrandt. There is no suggestion that a higher purpose of advancing scientific or medical knowledge is being served. The men attending the dissection vary in their responses. Some are bored or distracted, some appear over-eager or pompous, though none looks as animated, ironically, as the subject of investigation.

Dissection in Hogarth's London in the 1750s has a very different character to Rembrandt's Amsterdam anatomies of a century earlier. The difference between the two, however, is not a simple chronological progression. As we have seen, literary and dramatic representations of anatomy in the seventeenth century understand it predominantly as a macabre and grotesque practice, the social context of which is more likely to be a conjunction of a base criminal in need of the worst kind of punishment with an arrogant devotee of a ghoulish art. Thomas Nashe's picaresque *The Unfortunate Traveller* [1594] sees the hero, Jack Wilton, in a scene which is both revolting and comic, faced with the prospect of becoming 'an anatomy' and sliced open 'like a French summer doublet'. In his survey of early modern literary treatment of dissection and anatomy, Sugg notes that in the seventeenth century the vocabulary of cadavers is still plastic and unfamiliar[48]: the word 'skeleton' requires glossing, even in tracts with an educated readership, and 'an anatomy' might refer to a preparation, a skeleton or even to gibbeted remains.

INTO MODERNITY

Until at least the end of the eighteenth century, anatomy was 'intimately connected' to the criminal process.[49] Possibly it was not until after the Anatomy Act and the formal removal of the criminal corpse from anatomical attention that anatomy was widely acknowledged to be a useful and legitimate science, and its practitioners to be motivated by benevolent research rather than impious glee in gore. But in the nearly two centuries since the Murder Act was formally repealed, writers, artists and, latterly, filmmakers have continued to invoke the criminal corpse as an emotionally powerful vehicle upon which to make political, social or symbolic points.

In contemporary art, film and literature, it is almost impossible to represent bodies that suffer excruciating pain or humiliation unsympatheti-
cally: the suffering body itself is such a powerfully freighted image that it
can be used in fact as an emotional short-cut directly to the empathetic
heart of the audience. In modern film history, the criminal corpse rarely
features. Aggravated executions are significant motifs of some films, and
are always redemptive in character. This is obvious in the films of Mel
Gibson, such as *The Passion of the Christ*, and the Christian narrative is
appropriated in the construction of a romanticised nationalist myth in
Braveheart.[50] 'Braveheart' tells the story of William Wallace, leader of a
Scottish rebellion against the English in the thirteenth century, whose
eventual execution for treason was described in Chapter 2. There is little
subtlety in the political history of the film: the English are almost univer-
sally despicable, greedy, violent, cowardly and effeminate. The Scottish
nobles are self-interested and perfidious. However, all the ordinary folk
of Scotland are brave, honest, selfless, charming and handsome, most
of all Wallace himself, played by Mel Gibson who also directed the film.
His intelligence, attractiveness, superiority as both a soldier and a lover
are heavily played off against the effete, cruel English kings. Eventually,
Wallace is betrayed to the English and convicted of treason, which means
he is to be punished by hanging, drawing and quartering. Wallace faces
death with dignity and refuses to show penitence or declare allegiance to
the English king, even under torture. This has the effect of bringing the
originally hostile crowd around to his side as they recognise his strength
of character, the legitimacy of his cause and manliness of his conduct.
After Wallace's execution, during which he is consoled not by a medieval
vision of Christ or by religious sentiments, but by a very modern appa-
rition of his dead wife, a voiceover tells the audience that Wallace's body
was quartered and the parts sent to the rebellious towns of Scotland, but
that they failed to subdue the freedom-loving hearts of the Scots, instead
inspiring them to further resistance. The final scene shows Robert the
Bruce, who throughout the drama has been torn between serving the
narrow interests of his family and the greater glory of his nation, leading
the Scots to victory at Bannockburn.

The great distance between 'Braveheart' and known historical fact has
been widely described. Many have also felt uncomfortable with the violently nationalist and homophobic message of the film. However, it was
massively popular worldwide and its story and characters were appropriated by sports teams and commentators, political parties, tourist boards

and many others with goods to sell or ideologies to promote. Colin MacArthur reviews these appropriations.[51] 'Braveheart' consciously fits the death of Wallace to the crucifixion of Christ. During a painfully pro-tracted death sequence, Wallace is strapped to a cross, resists the temp-tation of a diabolical 'confessor' who presses him to deny his cause, and eventually expires with the vision of his lost love miraculously before his eyes, and the word 'freedom' on his lips.

Nine years later, Gibson returned to the source, and once again depicted a martyrdom, this time the paradigm of the suffering body: the crucifixion of Jesus Christ. *The Passion of the Christ* was an ambitious and cinematic filmic depiction of the last hours of Jesus's life. It has been crit-icised not only for being excessively bloody, but also for its antisemitism (as with the English in *Braveheart*, the Jews in *The Passion of the Christ* are ugly and unmanly). But again, it shows the difficulty of represent-ing physical suffering without exciting the sympathy and horror of the audience. Aside from Jesus Christ's own death and resurrection, even the thieves crucified alongside him—properly criminal corpses—are not moral lessons so much as objects of our compassion.

Popular Belief, Cultural Production and Punitive Force

Was anatomical dissection a thing of such horror because it was a pun-ishment for the worst of criminals? Or was it an extreme punishment because it was a thing of such horror? The answer is that it was both. The terror of being dissected was recursively produced alongside the shame and dread of criminal punishment in a complex and shifting dance of cultural meanings. Dissection was dreadful because it was humiliat-ing, and in that capacity belongs to the tradition of punishment by pub-lic shaming; but it was also terrifying in a less logical and more visceral sense because of the slippage in imagination between the scalpel in the cadaver and the knife in the living body. Being cut was always horrific. Sugg's book of 2007 is entitled *Murder after Death*, a title that expresses well the terrors of post-mortem dissection. Two factors made the idea of cutting into the body especially horrific in early modern England. The first is that, as Katharine Park has described in relation to an earlier period, the newly dead in northern Europe, unlike their counterparts in Italy, were considered to be in the process of dying. Rather than being instantly blown out like a candle flame, life seeped gradually away from the wholly alive until they finally became wholly dead, when all the flesh had rotted away and their social presence among the living had ebbed. This process could take a year or more. Second, in an age before anaesthetics or effective infection control, and when even simple surgery involved agonising pain and a good chance of dying anyway from septicaemia or blood loss, there was no context in which the scalpel in the skin was not terrifying.

As surgeon Edward May lamented in 1603, anatomy was greatly impeded because the common people of the country would not allow their bodies to be investigated because they believe 'their children or friends murdered after they are dead, if a surgeon should but pierce any part of their skin with a knife'.[52]

Cultural productions such as art and literature used the criminal corpse to evoke strong emotional reactions, which could be turned to comic effect or developed as horror. But the existence of cultural works around the criminal body also constituted part of what made it such a fearful, repulsive and powerful thing.

The deeply disturbing cultural resonance of post-mortem punishment informed, in a way that was rarely if ever articulated, contemporary anxieties about the ethical treatment of the dead body. The final chapter of this book will consider how the dread of something after death continued to affect public attitudes and policy long after the end of the Murder Act, as well as the way that new narratives were sometimes able to contest the meaning, and even the very definition, of a criminal corpse.

Notes

1. Shelley, M. (1823), *Frankenstein: Or, the Modern Prometheus* (London: Penguin Books), quote at p. 91.
2. See, Tarlow, S. (2007), *The Archaeology of Improvement in Britain 1750–1850* (Cambridge: Cambridge University Press).
3. See, the first essay of Robert Owen's (1817), *The Human Character: Preparatory to the Development of a Plan for Gradually Ameliorating the Condition of Mankind* (London: Printed for Longman, Hurst, Rees, Orme, and Brown).
4. See Driver, F. (1988), 'Moral Geographies: Social Science and the Urban Environment in Mid-Nineteenth Century England', *Transactions of the Institute of British Geographers*, Vol. 13, 275–287; Cowling, M. (1989), *The Artist as Anthropologist: The Representation of Type and Character in Victorian Art* (Cambridge: Cambridge University Press).

5. See, Sawday, J. (1995), *The Body Emblazoned: Dissection and the Human Body in Renaissance Culture* (London: Routledge).
6. The best extended consideration of Donne's interest in early modern anatomy remains Jonathan Sawday's (1995), *The Body Emblazoned: Dissection and the Human Body in Renaissance Culture* (London: Routledge).
7. See, Andrews, W. (1899), *Bygone Punishments* (London: W. Andrews & Company), pp. 51–52. It is possible that Grindrod's story is the original because it was the subject of a popular ballad that was published in 1855 in W. Harrison Ainsworth's *Ballads: Romantic, Fantastical and Humorous* (London: Routledge), and it is certainly plausible that variants of this pleasing story were attached to gibbets in other localities.
8. See, Percy, R., Timbs, J., and Limbird, J. (1830), *The Mirror of Literature, Amusement and Instruction: Vol. 15* (London: J. Limbird), pp. 210–213.
9. Ibid., quote at p. 211.
10. See, Chassaigne, P. (1999), 'Popular Representations of Crime: The Crime Broadside—A Subculture of Violence in Victorian Britain', *Crime, History & Societies*, Vol. 3, Issue 2, 23–55.
11. See, Hood, T. (1831), *The Dream of Eugene Aram, The Murderer* (London: Charles Tilt).
12. See, Bulwer Lytton, E. (1832), *Eugene Aram. A Tale* (London: Henry Colburn and Richard Bentley).
13. Wodehouse, P.G. (1916), *Jeeves Takes Charge*.
14. Dickens, C. (1861), *Great Expectations*.
15. All we know of the gibbet is that it once held a pirate. The relevant sentences read: 'On the edge of the river I could faintly make out the only two black things in all the prospect that seemed to be standing upright; one of these was the beacon by which the sailors steered,—like an unhooped cask upon a pole,—an ugly thing when you were near it; the other, a gibbet, with some chains hanging to it which had once held a pirate. The man [Magwich] was limping on towards this latter, as if he were the pirate come to life, and come down, and going back to hook himself up again.'
16. Dickens, C. (1859), *A Tale of Two Cities*.
17. See, 'Book Twelfth', in Wordsworth, W. *The Prelude; or, Growth of a Poet's Mind: An Autobiographical Poem*.
18. See, Wu, D. (2012), *Romanticism: An Anthology* (Oxford: Wiley-Blackwell, 4th Edition).
19. See, Kelley, T.M. (1988), *Wordsworth's Revisionary Aesthetics* (Cambridge: Cambridge University Press), p. 120.

20. Gwynplaine was later played by Conrad Veidt in a 1928 film adaptation of Hugo's novel, titled '*The Man Who Laughs*', and has been cited as the inspiration for the DC Comics character, The Joker.

21. John Aitken, known as Jack the Painter, was executed and hung in chains in 1777 for arson at Portsmouth docks, in a case that attracted a great deal of press attention.

22. One of these might have been Thomas Brett executed in 1789 for piracy and hung in chains at Dover Castle. Although Brett was one of a group of defendants all executed in relation to the same offence, he was the only one ordered to be hung in chains at Dover; the others were to be displayed along the Thames.

23. Hugo, V. (1870), *By Order of the King* (London: Bradbury, Evans & Co. English Translation), quote at p. 85.

24. Ibid., p. 84.

25. Ibid., quote at p. 87.

26. Ornithologically unlikely; maybe rooks were intended.

27. Hugo, V. (1870), *By Order of the King* (London: Bradbury, Evans & Co. English Translation), quote at p. 89.

28. Housman, A.E. Poem IX, *A Shropshire Lad*.

29. Sherwood (1822), *The History of the Fairchild Family* (London: J. Hatchard and Son).

30. See, Sugg, R. (2007), *Murder After Death: Literature and Anatomy in Early Modern England* (London: Cornell University Press), pp. 213–216.

31. See, Sawday, J. (1995), *The Body Emblazoned: Dissection and the Human Body in Renaissance Culture* (London: Routledge).

32. Pope, A. (1742), *The Works of Alexander Pope, Esq; Vol III. Part II* (London: R. Dodsley), quote at pp. 61–62.

33. Aries, P. (1981), *The Hour of Our Death* (Harmondsworth: Peregrine), p. 367.

34. See, Sugg, R. (2007), *Murder After Death: Literature and Anatomy in Early Modern England* (London: Cornell University Press); Davies, O. and Matteoni, F. (2017), *Executing Magic in the Modern Era: Criminal Bodies and the Gallows in Popular Medicine* (Palgrave Macmillan); Brittain, R.P. (1965), 'Cruentation: In Legal Medicine and in Literature', *Medical History*, Vol. 9, Issue 1, 82–88.

35. Brittain, R.P. (1965), 'Cruentation: In Legal Medicine and in Literature', *Medical History*, Vol. 9, Issue 1, 82–88.

36. In Act I, Scene II, the dead Henry VI's wounds begin to bleed when Gloucester (his murderer) enters the scene. See, Lull, J. ed. (1999), *King Richard III* (Cambridge: Cambridge University Press), p. 63.

37. Scott, W. (2001), *Fair Maid of Perth*, (Classic Books Company Edition), p. 89.

38. See, Brittain, R.P. (1965), 'Cruentation: In Legal Medicine and in Literature', *Medical History*, Vol. 9, Issue 1, 86.
39. See, Davies, O. and Matteoni, F. (2017), *Executing Magic in the Modern Era: Criminal Bodies and the Gallows in Popular Medicine* (Palgrave Macmillan).
40. See, McCorristine, S. (2017), *Interdisciplinary Perspectives on Mortality and Its Timings* (Palgrave Macmillan).
41. See, Scott, W. (1995), *The Antiquary* (Edinburgh Edition of the Waverley Novels, 3), p. 138.
42. See, Matteoni, F. (2016), 'The Criminal Corpse in Pieces', *Mortality*, Vol. 21, Issue 3, 198–209; Davies, O. and Matteoni, F. (2015), '"A Virtue Beyond All Medicine": The Hanged Man's Hand, Gallows Tradition and Healing in Eighteenth- and Nineteenth-Century England', *Social History of Medicine*, Vol. 28, Issue 4, 686–705; Davies, O. and Matteoni, F. (2017), *Executing Magic in the Modern Era: Criminal Bodies and the Gallows in Popular Medicine* (Palgrave Macmillan).
43. See, McCorristine, S. (2017), *Interdisciplinary Perspectives on Mortality and Its Timings* (Palgrave Macmillan).
44. See, Davies, O. and Matteoni, F. (2017), *Executing Magic in the Modern Era: Criminal Bodies and the Gallows in Popular Medicine* (Palgrave Macmillan).
45. See, Byrne, G. (2010), *Spiritualism and the Church of England, 1850–1939* (Woodbridge: The Boydell Press), p. 90.
46. See, Kemp, M. (2010), 'Style and Non-style in Anatomical Illustration: From Renaissance Humanism to Henry Gray', *Journal of Anatomy*, Vol. 216, Issue 2, 192–208.
47. See, Heckscher, W.S. (1958), *Rembrandt's Anatomy of Dr. Nicolaas Tulp, an Iconological Study* (New York: New York University Press).
48. See, Sugg, R. (2007), *Murder After Death: Literature and Anatomy in Early Modern England* (London: Cornell University Press), pp. 19–35.
49. See, Sawday, J. (1995), *The Body Emblazoned: Dissection and the Human Body in Renaissance Culture* (London: Routledge), p. 62.
50. See, *Passion of the Christ* (2004) [film], dir. Mel Gibson (USA: Icon Productions); *Braveheart* (1995) [film], dir. Mel Gibson (USA: Icon Productions).
51. McArthur, C. (2003), *Brigadoon, Braveheart and the Scots: Scotland in Hollywood Cinema* (London: I.B. Tauris & Co.).
52. Edward May *Most certain and true relation*, cited in Sugg, R. (2007), *Murder After Death: Literature and Anatomy in Early Modern England* (London: Cornell University Press), p. 20.

CHAPTER 9

Conclusions: Ethics, Bullet Points and Other Ways of Telling

THE ETHICAL LEGACY OF THE CRIMINAL CORPSE

One of the project's key questions is 'How far have beliefs about the dead body, harm and criminality remained the same during historical journeys from sacred to secular, and from "ancien régime" to "modern" styles of justice?' In order to think about the diachronic aspects of this question, our larger project included a strand on the relationship of contemporary ethical anxieties about the treatment of the dead body to the attitudes revealed in the historical studies. Philosopher Floris Tomasini was focused particularly on the ethical dimension of the treatment of humans after death, and the idea of post-mortem harm.

As Tomasini describes, philosophical approaches to post-mortem harm have been broadly of two camps: in the first camp are those who reject the possibility that post-mortem harm is possible, a position exemplified in ancient philosophy by Epicurus, and in modern philosophy by Ernest Partridge.[1] For harm to be done to a subject, they maintain, it is a basic condition that the subject exist at the time of the harm. Dead people do not, they say, exist, and therefore it is a logical impossibility to do them harm. The second camp holds that a subject's interests can be harmed after their death. This approach, given its most sophisticated expression in the work of Feinberg and Pitcher,[2] develops an argument that the ante-mortem interests of the subject can be retrospectively harmed by an act which, for example, fails to respect their body or their

S. Tarlow and E. Battell Lowman, *Harnessing the Power of the Criminal Corpse*, Palgrave Historical Studies in the Criminal Corpse and its Afterlife, https://doi.org/10.1007/978-3-319-77908-9_9

wishes after that subject's death. So if we should choose to ignore our friend Helen's great fear of her body being burned and arrange for her corpse to be cremated, we do harm to the living Helen. Similarly, if we arrange for her estate to be donated to a political party to whose policies Helen was opposed, we harm her interests.

Tomasini recognises the great contribution made by this second position in clarifying that a person's social existence is not co-terminous with their biological life. He points out that people have a narrative identity as well as a biological one (a body), and that social death is not the same as medical death. Social death is, rather, 'a relational or narrative change in the meaning of a human life … a change in the narrative identity of persons that either still exist or have once existed'.[3] Because social/narrative identities live on after the point of an individual's biological death, individuals therefore have transcendent interests that outlive them. For Tomasini, the time-travelling contortions of Feinberg and Pitcher are not necessary. Instead one simply rejects the Epicurean assertion that the dead do not exist. Life is not the same as existence. The interests of a relational or narrative self can still thus be furthered or harmed by posthumous events.

Our research into post-mortem punishment in the eighteenth and nineteenth centuries in Britain shows that people of the time clearly believed in the possibility of post-mortem harm. We are, of course, cautious about positing that a belief in the possibility of post-mortem harm is universal. The ubiquity of archaeological, historical and ethnographic examples of punishing the corpse, however, make it at least widespread in actual material practice. The punitive treatments of the deviant dead in the early medieval period, as outlined in Chapter 2, are examples of this. The dead body could be a site of shame and humiliation as well as celebration, veneration and glamour. For this reason, the story of an individual obviously does not end with their death; the individual continues to be represented into the future, and, as the subject of representation, clearly does have interests that can be damaged or promoted through that representation. The corpse is an important material resource in the process of representation.

Tomasini develops his ideas through a consideration not only of the post-mortem harms done to executed criminals in our main period of study, but also through the twentieth century examples of the British soldiers of the First World War, who were shot at dawn by their own side for desertion or cowardice, and the organ retention scandal at Alder Hey

Hospital, Liverpool, when it was discovered that organs of dead babies and children had been kept without the knowledge or consent of their parents in the 1980s and 1990s. In that case, the distress caused to the bereaved families when they discovered that what they had buried was not the whole child but 'a husk' was almost as great as that they had suffered at the time of the child's death. The ensuing outcry actually precipitated a change in the law, and helped to crystallise best practice in contemporary medical ethics.[4]

The case of the 'shot at dawn' soldiers is a fascinating one, and gets to the root of the fundamental question 'what is a criminal corpse?' During the First World War, around 3000 people were found guilty of capital crimes by courts martial—courts staffed and convened by the armed forces outwith the normal judiciary of Britain, but with special powers, including sentencing and execution. Of those sentenced to death by courts martial, around 90% had their sentences commuted, but 346 people were executed by their commanding officers and their comrades. These convictions break down as follows:

During 4 August 1914 to October 1918 there were approximately 238,000 courts martial resulting in 3080 death sentences. Of these only 346 were carried out, which break down into the following categories of offences on active service:
Mutiny 3
Desertion 266
Cowardice 18
Disobedience of a lawful order 5
Sleeping at post 2
Striking a superior officer 6
Casting away arms 2
Quitting post 7
Murder 37.[5]

Execution was typically carried out by a firing squad comprising members of the condemned man's own regiment. To be shot at dawn was a shameful death. The names of those so executed were not included on war memorials, and the shame of their death frequently caused ongoing and additional stress and distress to their families, sometimes for many generations.[6] It is undoubtedly the case that many of the men found guilty of desertion or cowardice were suffering from what would now be recognised as post-traumatic shock and were not in a mentally responsible state.

Many of the convicted would not even be considered capable of standing trial in a modern court. But at the time of their deaths, psychological understanding of the effects of war was not well developed, and legal culpability was assumed. The decision to execute rather than enforce an alternative punishment was inconsistent and often seemed arbitrary. Personal relationships between the convicted man and his senior officers were very significant, as was the fluctuating need to make an example. The effects of class were evident, in that very few of the men executed were from the senior ranks or the middle classes (often the same thing).

Increasing attention to the fates of those shot at dawn from the late twentieth century eventually led to the decision in 2006 by the (then) Defence Secretary, Des Browne, to issue a blanket pardon to 306 people shot at dawn (so excluding those executed for murder or mutiny). Reactions to this decision were varied and complex. Supporters of the decision to pardon saw the issue as one of righting a historical injustice, acknowledging and mitigating the harm done to families and descendants (Fig. 9.1). On the other side were those who felt that a pardon nearly a century later was anachronistic and 'rewriting history'. It is inappropriate, they claimed, to judge the actions of people in the past by the standards and with the knowledge of today. Within the context of their time and place, the judgements made were reasonable. Moreover, if one person, or group is selected for a retrospective pardon, then justice surely demands that every historical conviction and punishment be similarly reassessed, which is nearly impossible at this distance, and not the best use of judicial time or energy. Above all, said the critics, there is no point in issuing a pardon now. The damage is done.

The issue of posthumous pardoning illuminates a tension facing historians, archaeologists and anyone attempting to write about the past, to tell a story of what happened and make some kind of narrative sense of events. On the one hand, traditional historians are anxious that their value of fidelity to the past may be undermined by the kind of anachronistic engagement represented by the posthumous pardon, which appears to neglect historical context. On the other, postmodern historical approaches that arose in the second half of the twentieth century acknowledge that 'history' is not immutable and is 'an unending dialogue between past and present'.[7] The posthumous pardon recognises that the family narrative is also valid and that a new history, which is only ever provisional and partial anyway, will be informed by new knowledge and changing moral codes.

Fig. 9.1 *Shot at Dawn* memorial, National Memorial Arboretum (Sarah Tarlow)

If it is sometimes contentious that changing knowledge and social values in the present can or should change the kinds of pasts we write, it is much more generally acknowledged that narratives and values created in the past in a particular set of historical circumstances will shape the terms of contemporary debate. One of the interesting aspects of our work on the criminal corpse is the way that eighteenth- and nineteenth-century attitudes to the dead body continue to colour often unspoken beliefs about death and the body into the modern era. As is the case with many punishments, the formalised use of a particular treatment as a punishment acts back on the sanction to make what might otherwise be a morally neutral treatment a humiliating and punitive one. This is what has happened in the case of anatomical dissection. While cutting the corpse was undoubtedly already distasteful, at least in northern Europe, by the Renaissance,[8] the use of dissection as a punishment for the most serious crimes, and its association with a context of public humiliation strengthened the general view that to have one's body cut after death was a deeply distasteful and shaming fate.

STUDYING THE CRIMINAL CORPSE: OUR OWN ETHICAL POSITION

There is, of course, an accusation of ethical culpability that could be levelled at our project as a whole. We have spent five years, and produced dozens of publications discussing the fate of criminal bodies. Their post-mortem treatment was frequently brutal, vengeful and pitiable. In many of the stories we have told and retold, the criminal body at the core assumes the part of victim; the villains of the drama are, implicitly, the legislators, the sheriffs, magistrates, judges, surgeons and sometimes crowds of the vengeful but unspecific 'public'. In our work, we have brought the names of executed criminals back into mind and arguably reinforced their entitlement to be considered important historical actors. We have aided in their remembrance and, while we have tried not to romanticise these individuals, a historical review like ours has demonstrated the difficulty of finding a narrative of individual punishment that does not bear traces of heroic story.

And yet those men and women whose bodies were opened or displayed were not—or not only—plucky Davids facing the Goliaths of Law, Science and the State. They were certainly not sacrificial Christs subject to the arbitrary cruelty of an unequal power struggle. They were convicted murderers. Often their murders were violent and the true victims were frequently very young or very old, and relatively helpless. The people we have studied killed for the most despicable of motives: greed, lust, uncontrolled anger or envy. They were not heroes and do not deserve to be remembered or commemorated. Perhaps they more properly merit a *damnatio memoriae* in the Roman sense: to have their names chiselled off monuments and excised from records. Instead we have published articles about them and produced public lectures and educational websites.

Anyone who writes about killers, terrorists, or criminals necessarily walks a line between analysis on the one hand; and on the other the 'oxygen of publicity'.

Barry Godfrey proceeds from the position that only research that has a directly detrimental effect on the living can ever be unethical, and therefore that 'For the most part, historical research need not trouble the ethics panels'.[9] However, many historians, ethicists and archaeologists are not satisfied with this get-out-of-jail-free card, and ask instead to which other groups, individuals or even principles we owe an ethical duty. These might include descendants of both direct genetic lineage and communal identity, students, funders and the people of the present day.[10] Surely nobody is so naïve as to suggest that the work of historians cannot be turned to advantage by those pursuing political ends, including emancipatory, nationalist, liberal and conservative agendas. It would be disingenuous to maintain that historians have no responsibility for the way their work might be exploited in buttressing conclusions that are not their own.[11] However, ethical responsibility arguably also extends beyond our contemporaries to both the people of the future and to those past people about whom we write.[12] Sarah Tarlow has argued elsewhere that our responsibility to the people of the past should be understand as a duty of representation, a responsibility that is

> perhaps closer to 'justice' or 'fairness' than 'truth'. Although there is no right way to represent people of the past, there are wrong ones. These right and wrong forms of representation are unlikely to remain constant, however, and their moral imports will be decided by factors including their likely social and political effects in the present. Many forms of just representation will be possible, and understandings of what is 'justice' are neither constant nor transcendent... One interpretation of 'just representation' (but by no means the only one), involves finding ways of representing the people of the past which emphasise some of the richness and texture of their experiences and gives weight, where this is possible, to some of the values and understandings by which they understood their own world.[13]

We have not adequately resolved this conundrum, but felt increasingly troubled by it as the project progressed. In future, research and analysis focussing more on the names and stories of the victims of violent crime may help to redress this imbalance.

What has become clear through the work of the group, and especially through Tomasini's consideration of the contemporary ethical implications of historical research on the criminal corpse, is the impossibility of developing proper ethical practice in a context-free present, which takes no account of the deep history of the body. In the opening pages of her seminal study of the social history of the Anatomy Act, Ruth Richardson observed that even in the 1980s her older neighbours had a terror of receiving a pauper's burial.[14] Although the social stigma of being a recipient of parish welfare might have played a part in this, Richardson attributes it to a collective folk memory of the time when dying 'unclaimed' in the workhouse meant that one's body would be taken for anatomical dissection and would not be buried at once or intact.[15] It is undoubtedly the case that even now many people find the idea of a human body being used for the research or education of medical and biological scientists to be disturbing or even horrifying. Anxiety about such a fate is not wholly rational.

But fears and feelings about the fate of the body are neither generated nor addressed through rational evaluation of philosophical propositions, nor are they resolved by scientific facts. The parents of the Alder Hey children, like those who have protested about the exhibition of Gunther von Hagens's plastinated cadavers in his popular 'Bodyworlds' exhibition, participate in a deeper history, and draw on collective and historical memories and belief systems wherein intervention in the dead body for scientific purposes is considered to be a violation. This distrust of scientific cutting exists notwithstanding that intervention in the dead body for the purposes of embalming is common and usually attracts no protest.

A deep history of cultural attitudes to the treatment of the dead can lead to two different interpretations: either that distaste for interference in the body of a person who has died is a universal human attribute, working at a visceral level which defies logical explanation, or it is the product of particular and contingent histories. Both authors of this volume having an anthropological background and bent, incline towards the latter position. The anthropological literature is replete with ethnographies of death that demonstrate the range and diversity of cultural responses to the universal fact of death, including a wealth of treatments of the dead body itself. What constitutes 'normal', 'respectful', 'disgusting', or 'decent' cannot be glossed in cross-cultural perspective. In their edited volume, Metcalf and Huntingdon bring together examples of dead bodies being buried, burned, pickled, exposed and absorbed into a tree.[16] We could add examples of cultures in which respectful treatments of the dead body include eating it, embalming it, keeping it in the family home, freezing it, exhibiting it, sinking it in the sea or blasting it into space.

The notion that a dead body should be quietly buried, shielded from view and left undisturbed is a historically specific one. Among the Andean Inca, for example, great leaders contrived to be socially active long after their biological life, not only owning property, but also physically participating in ceremonies, processions and feasts, as their embalmed bodies were paraded through the streets and given food and beer.[17] To this day, the Merina of Madagascar regularly remove the remains of their ancestors from the collective tomb whenever a new burial takes place, so that the living may dance with the dead, before the bones are ceremonially rewrapped and replaced inside the tomb.[18] There is nothing natural or inevitable about the north European tradition of burying the bodies of the dead intact and undisturbed. Indeed, this is particularly evident in the recent and striking change to practices of respectfully disposing of the dead in the United Kingdom. In 1900, over 99.9% of those who died in the United Kingdom were buried, but in 2014, nearly 75% were instead cremated—a remarkable shift in a relatively short time.[19]

The post-mortem punishments of dissection and gibbeting only work in a historical context where such treatments outrage the norms of disposal. The provisions of the Murder Act permit sanctions that only work in contexts where anatomisation or hanging in chains are already horrifying, because of their particular histories and traditions. Twentieth- and twenty-first century ethical anxieties about the treatment of the dead in Britain, as studied by Tomasini, partake of those same histories.[20] Normative cultural practices shape attitudes towards the dead body, just as attitudes towards the dead shape normal (and exceptional) cultural practices. The relationship between practices and feelings is recursive.

Stories We Could Tell About the Criminal Corpse

How can one, how should one, talk about the criminal corpse? It is possible to identify many conceivable narratives about the history of the criminal corpse. These are not necessarily incompatible, but emphasise different aspects of the post-mortem treatment of the executed criminal. The list that follows is not exhaustive.

1. Marxist
   The spectacular display of suffering and humiliation visited on the deviant body can easily be read as an emphatic demonstration of state power, designed to prevent the oppressed proletariat from challenging the established order by impressing upon them the consequences of social deviancy.[21] More subtly, the theatricality

of the public dissection or carnival of gibbeting can be read as 'bread and circuses': a spectacle which recruits the problematic and ambivalent 'crowd' to become complicit in a celebration of social conformity and an 'othering' of deviance.[22] In the body of the individual criminal, ostentatious and public post-mortem punishment creates a scapegoat for society's problems, which distracts attention from the deeper structural inequalities which are an essential part of the emergence of a parasitic capitalist class.

2. Part of the History of the Body

Using post-mortem anatomical dissection as both a legal sanction and a research practice represents the intersection of two histories of the body: the first is the body as a site of punishment and legal control, and the second is as a place of expanding scientific knowledge. The Murder Act represents a particular stage in the evolving relationship between self and body. The growing anatomical and scientific understanding of a universal medical body is fundamental to the growth of modern medicine as a practice based on empirical observation, experimental and replicable science, and contrasts with a premodern medicine based on ancient authority and divine grace.[23] Its public nature is evidence of a technology of learning through which a scientific understanding could be expanded and democratically shared, and the place of the dissected corpse within the society of medical men is important in the history of medical education. Formal and informal pedagogical structures allow the cadaver of the executed person to be used to improve personal familiarity with the body's interior. Despite the very small number of bodies coming into the hands of medical men from the scaffold, in comparison with those acquired from relatives, sextons, grave robbers and by other unofficial channels, the legitimate and predictable acquisition of criminal bodies meant that they could be used in more public, planned and sanctioned ways.[24]

3. A Ghoulish Horror Story

While writing this chapter, we were asked by our employing university's press office to produce a story for Halloween release.[25] Gibbeting, human dissection and capital punishment are still considered both ghoulish and thrilling: a frisson of fright and the pleasure of the grotesque, but no real danger. The swinging gibbet continues to be used as an atmospheric bit of scenery in modern films and television plays.[26]

4. Feminist
   Because the capital crime of murder was much more likely to be committed by men than by women, the criminal corpses we study are overwhelmingly male bodies. This meant that on the rare occasions that a female body came within the scope of the Murder Act, it was a scarce commodity, much sought after especially by the surgeons. This is likely to be the reason that our project did not find a single woman among the records of those whose bodies were hung in chains in Britain during the life of the Murder Act: all were requested for anatomical dissection. But the desire to explore a female body was not only motivated by a need to remedy an imbalance in research material. Michael Sappol has discussed the sexuality of anatomical dissection.[27] Only in anatomy and fine art, he says, was the display of naked female flesh acceptable. Examining works of popular fiction, Sappol notes the sensationalist or even pornographic depiction of the penetration of female flesh by the anatomist's knife and the masculine scientific gaze.[28] Helen Macdonald notes the artistic depiction of medical men lasciviously ogling the undefended flesh of a female cadaver (Fig. 9.2).[29]

5. A Tragedy (1)
   Disrespectful treatment of the dead body is a classical motif at the heart of Sophocles's classical tragedy *Antigone*, and features in many other classical stories, especially ancient Greek ones where depriving a body of proper funerary rites was among the worst of offences. Antigone's struggles to come to terms with the death of her brother are made far more agonising by Creon, the king of Thebes, who orders that his body should be left unburied as food for worms and birds. The brother, Polyneices, has been slain in a civil war against his own brother and the new ruler has decided that as a punishment for leading foreign troops to his own city, Polyneices's body should remain unburied while his brother Eteocles should be buried with honour. Antigone cannot bear this and argues that a higher law than the king's—the law of the gods, demands that Polyneices's body be buried. Unable to persuade anyone to do this or to help her, she buries the body herself. The rest of the story is then occupied with the fate of Antigone, who is sentenced to death by Creon for putting the honour of her dead brother and the will of the gods above the will of the ruler who personifies the interests of the state.[30] In the classical world, the proper disposal of a dead body was of supreme importance,

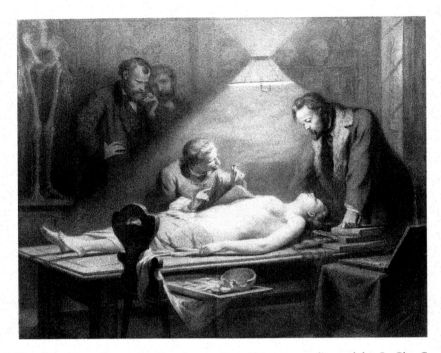

Fig. 9.2 The dissection of a young, beautiful woman directed by J. Ch. G. Lucae (1814–1885) in order to determine the ideal female proportions. Chalk drawing by J. H. Hasselhorst, 1864 (Wellcome Collection)

and its dishonourable treatment was the worst offence. The tragedy of Antigone works by engaging the sympathy of the audience for the heroine, whose grief in bereavement is made worse by her inability to give him proper funerary honours. Even in the very different context of eighteenth-century Britain post-mortem shame of the body had the power to engage the sympathy of onlookers for the person punished and their family, rather than securing their identification and alignment with the forces of justice and law. Newspaper and periodical accounts of gibbettings sometimes include poignant detail of the visit of a parent to the foot of the gibbet: In the case of the Drewitt brothers, who were hung in chains in 1799 in Sussex, the boys' father 'spent the remainder of his days in sitting at the foot of the gibbet on which swung the bodies of his two sons'.[31] Returning to the case of Spence Broughton from Chapter 6, by 1900 his story included not only the piteous letter of love and repentance he supposedly wrote to his estranged wife on the eve of his execution, but also a poignant vignette: the widowed Mrs. Broughton sitting alone in the window of the Arrow Inn with 'tear-dimmed eyes' watching her husband's body 'swinging there 'twixt heaven and earth'.[32] Whether factual or not, the image of the noble yet bereaved woman bearing solitary witness to the decay of the body of the man who was once her husband tugs at the heartstrings (even of the modern reader), and embeds a sense of tragic romance in the history of a notorious criminal.

Fanciful, fictional accounts of the pathetic meditations of the bereaved relatives and lovers of the executed increased during the nineteenth century. Bulwer-Lytton's *Eugene Aram* would be just such a text. The deservedly little-known poet William Newton was inspired by Anthony Lingard's gibbet to compose an ode entitled '*The supposed Soliloquy of a Father, under the Gibbet of his Son; upon one of the Peak Mountains*'

TIME — Midnight. SCENE — A Storm.
[Naturally. And the poem ends]

...Art thou, my Son, suspended here on high? —
Ah! what a sight to meet a Father's eye!
To see what most I prized, what most I loved.
What most I cherish'd, — and once most approved,
Hung in mid air to feast the nauseous worm.
And waving horrid in the midnight storm!
...— When heretofore
Our barbarous sires the aweful Gibbet rear'd.
The Gibbet only, not the laws were fear'd:
The untutored ruffian, of an untaught clime,
Fear'd more the punishment than dreaded crime.
We boast refinement, say our laws are mild.
Dealt equally to all, the man, the child: —
But ye, who, argue thus, come here and see,
Feel with a Father's feelings; — feel with me!
See that poor shrivell'd form the tempest brave.
See the red lightning strike, the waters lave.
The thunders volleying on that fenceless breast! —
Who can see this, and wish him not at rest?
...
O! blind to truth, to all experience blind.
Who think such spectacles improve mankind:

Sentimental details such as this account undermine any attempt to make a narrative in which the gibbet represents the ultimate triumph of good over evil: there is clearly no happy ending for the innocent and vulnerable elderly people whose old age is now blighted by grief, shame and probably material want as they can no longer expect to be supported by their children. Newton's poem uses a sentimental appeal to the reader's empathy to undermine the justice of the state, which is callous and cold.

6. As Tragedy (2)

A tragedy involves the ruinous downfall of an otherwise honourable protagonist because of a character flaw or an ill-judged decision. In this kind of tragedy, the audience's sympathy is mainly with the criminal who is executed and then subject to post-mortem punishment. His or her victims are pushed into the background, and the murderer is recast as a victim on their own account. They are the heroes of their own stories: clever, brave, maverick. These kinds of stories glamorise the criminal and, while they can draw attention to the inhumanity of capital punishment and its aftermath, there are ethical implications in focussing on what the criminal had to endure rather than on the suffering of his victims, as was discussed earlier. Bulwer Lytton's sentimental Eugene Aram, for example, was a tragedy of this kind.

7. As a Political Expedient

Although the wording of the Murder Act proclaims its purpose to be the better prevention of the horrid crime of murder, by the middle of the eighteenth century the incidence of murder was already in decline,[33] and there had been no particular epidemic of killing in the period leading up to the act. However, there had been a moral panic in the press, in response to a small number of high profile cases near London, which might have given rise to the erroneous perception that murder was becoming more prevalent. Whatever the truth of the matter, a widespread popular belief that people were in greater danger demanded a political response. The Murder Act was a visible response by a government that needed to be seen to be coming down hard on violent crime. In this argument the efficacy of dissection or hanging in chains as a deterrent to the commission of murder is less important than its efficacy in demonstrating that the government would not tolerate violent crime and would offer a muscular and decisive response to allay public fears, while demonstrating its own authority and puissance. This is a classic 'moral panic', as described by Peter King and Clive Emsley in which sensational reporting whips up public anxieties which are eventually calmed by the authorities' response, often one of harsher legislation, and the passage of time.[34] If the purpose of the Murder Act was for the government to be seen to be doing *something*, then it had the further advantage of necessitating repeated public displays of the State's resolve. Each iteration of post-mortem corporal punishment acted as a mnemonic of the Act and a further demonstration that the government was taking action to reduce or eradicate violent crime. Whether it actually worked is not the point. Like gassing badgers or leaving the European Union, taking visible and resolute action mattered to the British government more than taking effective action.

8. As a Successful or a Failed Experiment

Did the Murder Act work? Did it in fact add 'some further Terror and peculiar Mark of Infamy' to the punishment of execution? It does appear that for many condemned men and women the dread of having their body anatomised or hung in chains after their death was a significant additional terror. A number of felons begged to have that part of their sentence remitted, or openly bemoaned the fate of their bodies.

Did it better prevent the horrid crime of murder? That is harder to assess. As King has recently summed up, at the time of committing murder either a belief in one's ability to avoid detection, or an emotional state sufficiently pronounced as to occlude rational judgement probably meant that a balanced consideration of the likely post-execution consequences of crimes probably did not play a role in the criminal's decision-making process before the Act.[35] Even if it did, the sanction of death was surely enough on its own to stay the hand of any murderer likely to be swayed by such considerations, and, as the *Leicester Chronicle* asked in 1832 'If the terrors of a violent death cannot deter the murderer, will the dread of having a few incisions drawn upon his lifeless and unfeeling corpse wield a greater influence?'[36]

Post-mortem punishment was not a practice limited only to Britain. As discussed in Chapter 1, corpses have been the subject of harm and exclusion from burial rites to punish the living and the dead in many parts of the world. However, over the course of this project, we became curious: was there anything particularly

'British' about the post-mortem punishments mandated under the Murder Act? The life of the Murder Act encompasses an important period in the history of British imperialism and colonisation. During this time, Britain expanded its overseas empire aggressively in the Americas, Australasia and the Indian subcontinent. One way to respond to this question is to ask whether anatomisation and dissection and gibbeting were part of the suite of techniques and technologies transmitted or transplanted to the colonies as part of the legal and cultural spread of the British Empire.

We know that gibbeting was a form of punishment used in addition to hanging to punish murderers in Australia, Canada, America and India. Extant gibbet cages from some of those locations in addition to textual evidence demonstrates that in these places gibbeting closely followed the form evident in Britain. As discussed in Chapter 6, we also know that gibbeting was used in the plantation colonies as a much more vicious form of prolonged torture, execution and post-mortem display of enslaved black individuals. The use of gibbeting as a punishment for murderers in the overseas British world ended not long after the end of the Murder Act. The last known instance occurred in 1837, when John McKay was gibbeted at the site of his murder of Joseph Edward Wilson, near Perth Australia.[37] So, was anatomisation and dissection similarly in evidence as a post-mortem punishment in the British world in this period?

As Clare Anderson has found, dissection was practiced on criminal corpses in overseas territories administered by Britain's Colonial Office.[38] However, in the main this does not seem to have followed the form of 'anatomisation and dissection' as practiced in Britain, as it lacked the public demonstration aspect. This opportunistic use of criminal corpses for medical dissection occurred not just on land, but also on water. As Katherine Foxhall has identified, the corpses of some British convicts who died in the course of their transportation to Australia were dissected by shipboard surgeons.[39] In these cases, dissection was an extremely private affair as should the practice become known to the ship's population, it could provoke objections that might become dangerous to the ship's safety. Surgeons dissected in these situations in the interest of their own training and investigation, not as part of a demonstration of state power or additional sanction. Nonetheless, the fate of these criminal corpses continued the connection between criminality and dissection formalised by the Murder Act.

The prison hulks were created following the interruption of convict transportation by the American Revolution and used as holding cells in Britain for those later destined for transportation overseas, but also to house convicts sentenced to hard labour who were put to work carrying out colonial labour in Bermuda (1824), Ireland (1826) and Gibraltar (1842). The hulks were also a source of criminal corpses for dissection. In the early days of their use, death rates on the hulks were 'appalling' even compared to other prisons of the period.[40] Of the 632 prisoners incarcerated on the hulks from August 1776 to April 1778, 176 died. This rate—approximately 1 in 4—held steady for the first 20 years the hulks were in use, with a total death toll during this period of about 2000.[41] These corpses were buried along the banks of the waters where the hulks were moored, often in shallow sandy graves. However, an unknown number were 'sent for dissection, a side line which, according to one former prisoner, earned the hulk doctors £5 or £6 a corpse'.[42] Again, the connection between criminality and dissection continues, even in the absence of the formalised punishment of anatomisation and dissection.

Certainly, gibbeting was part of the suite of punishments Britain transported to the colonies as part of imperial expansion during the period of the Murder Act, and in cases involving white British (overseas) citizens, was carried out following the form and processes used in Britain. Punitive dissection that follows the way this punishment was conducted under the Murder Act in Britain, however, has been more difficult to identify. Nonetheless, the widespread dissection of criminal corpses on land and water in the British Empire served to further the connection between criminality and dissection in this period.

FINAL CONCLUSIONS

1. First, the journey of the criminal continues beyond the gallows. Peter King developed the notion of the criminal journey as a useful metaphor to understand the processes of decision making and discretion between apprehension of a criminal and their eventual fate: execution, transportation, some form of corporeal or financial penalty, the deprivation of liberty or exoneration and freedom.[43] King conceives the journey as a progression through a series of rooms, each of which leads to different possible spaces depending on the

outcome of the decision made in it. What our project has done is to extend that journey beyond what looks like the last room: the one with the noose, the stake or the axe in it. Even in contemporary accounts such as newspaper reports, pamphlets and ballads, most criminal stories finish when the malefactor is 'launched into eternity' on the scaffold. In fact, they were only launched into the next phase of their own biography. Post-mortem criminal histories build on key continuities with what went before and are therefore legitimate extensions of individual historical narratives. To the tradition of biography and life writing must be added relevant death writing and individual necrographies.

Narrative post-mortem histories are both personal and collective. Individual and unique lives were transformed on the gallows and went on to become individual and unique afterlives, in which the notoriety of the criminal, their glamour,[44] and often the web of emotional relationships in which they were embedded continued to shape the experience of those around them. Modern psychological approaches to death and bereavement highlight the importance of continuing bonds: the ongoing capacity of the dead person to affect those left behind, and the relationships between living and dead that extend beyond the moment of separation. Such relationships might be characterised by love and grief, but could also be relationships of hatred, anger, fear, envy or any number of complex emotions.[45] The criminal corpses in our study might be looked at, spoken to, made the butt of jokes, the object of fear, or the theme of a moral lesson for children. They might be transformed into landmarks, research data or teaching aids. They might be used to prove a theory or cure a disease. Their bodies might occasion anger or grief in those still living.

At the same time, the criminal dead participate also in a collective identity, as generic and deindividualised examples of a type. That type might be a universal medical body representative of a certain age and sex, or it might be as a member of the general category 'murderer'. Criminal corpses have both a practical importance and a symbolic or emblematic one. The practical one is primarily significant in scientific and medical fields where bodies were important to medical education and research, and could be mined if particular parts were needed for special study, or as components in folk medicine. For these purposes the name and particular life history of the individual to whom the body belonged was not important.

2. The history of the criminal corpse is widely present in the contemporary world, but is not well understood. There are many places around the country that still bear the name of the person whose dead body was displayed there, but there is little remaining folk memory of the significance of names such as Old Parr Road (Banbury), Tom Otter's Lane or Toby's Walks. When Tarlow made an appeal on national radio in June 2012 for information about the former locations of gibbets, none of those who got in touch identified those sites. However, numerous people knew of places called Gibbet Hill, Gibbet Woods, Gibbet Lane and so on. In fact, most of these generic gibbet place names predate the Murder Act, sometimes by several centuries. There are two kinds of historical slippage at work here: first there is the slippage between the gallows on which executions were carried out and the gibbets on which the bodies of the already dead were exhibited. Many 'gibbet' place names refer to executions that formerly occurred there. Second, there is an anachronistic compression of many centuries into a generalised past. The first of these is in some ways a reasonable elision since, as Poole has pointed out, it was common in some parts of the country for both execution and display to be carried out at the scene of crime.[46] However, by the eighteenth century it was more usual for the condemned to be executed at a customary place of execution, often a permanent gallows erected in a prominent urban location, and then removed for enclosure in irons and transportation to a specified place near the scene of the crime for gibbeting. The confusion between gallows or scaffold on one hand, and gibbets on the other is only one popular confusion about the nature of post-execution punishment in Britain. Contrary to widespread belief, nobody in Britain during this period was sentenced to be dissected or gibbeted while still alive. The punishment in both cases was in the apprehension by the condemned of the fate of their body after death, not in consciously experiencing the anatomist's knife or in looking out at the world from within a gibbet cage.

   The chronological confusion about when the age of post-execution punishment actually was is both distanciating and dehistoricising. It is surprising that less than two hundred years ago it was still legally mandated that a murderer's body should be mutilated or humiliated by the state. Failing to distinguish between medieval

Gibbet Woods and early nineteenth-century places of ritualised display allows post-execution punishments to be located safely in a rather foggy 'long ago'. It thus becomes possible to represent the sanctions of hanging in chains or anatomical dissection of the dead body as grotesquely humorous, in a way that would not be possible were the bodies in question thought of as more recent, their history rawer or their individual life stories and circumstances acknowledged.

However, in another way, it would be wrong to draw too sharp a distinction between the past and the present. If modern consumers now find it acceptable to use a cheap and ugly carica-tured model of a gibbeted body as a creepy, but funny, piece of Halloween décor rather than an awe-inspiring demonstration of the power of the State and the implacability of Justice, so too did their eighteenth-century forebears (Fig. 9.3). Even at the time of the Murder Act, many of those viewing the dissection in pro-gress, or the suspended corpse found it a subject for jokes as well

Fig. 9.3   Halloween decoration of a gibbeted criminal, on sale in a British supermarket (Sarah Tarlow)

as the source of scary thrills. The mismatch between what the state intended by post-execution punishment and what it actually achieved will be considered below.

3. A small number of bodies had a large social impact. Over the life of the Murder Act approximately 923 executed criminal bodies were sent to be anatomically dissected as part of their sentence, and only 144 were ordered to be hung in chains. There was considerable local and regional variation in the frequency with which post-execution punishments were carried out, with higher numbers in London, the Home Counties and in some eastern and southern areas, and generally a lower frequency in the north and west,[47] even taking into account the distribution of the population in general. However, the impact of each event was high. The number of witnesses to a gibbeting or dissection was maximised by, in the case of hanging in chains, the careful choice of conspicuous, open locations, which would permit large crowds to assemble and get a good view. Crowds of 10,000 people or more were often reported in the newspapers. The numbers able to view a dissection were limited by the need to use enclosed, internal space, but a number of strategies were developed to increase the number of ordinary people with direct visual access to the body. These included displaying the corpse, either straight after hanging or after initial incisions had been made, in a public open space in front of the building in which the full dissection was to be performed; controlling the movement of the crowd so that large numbers of people could file past the body as it was laid out; and ensuring that the dissection was carried out over a period of several days, allowing ticketed access to different groups in society (e.g., better-off people, women, men of science) access to the body at different times and different stages of the process. Even years after the dissection or enclosure in irons, the material remains of the body frequently endured in a visitable place and condition, either in their original landscape settings, or as part of museums and medical exhibitions (Chapter 7). These criminal bodies became their own mnemonics.

But the stories of these notorious criminals and their grim ends were also perpetuated through stories—mostly in the form of pamphlets and ballads. Some caught the popular imagination and inspired literary afterlives of a more enduring kind: Eugene Aram, whose story inspired a novel and a popular narrative poem, is our particular favourite; William Corder's titillating tale of love, betrayal, murder and eventual detection and comeuppance, which spawned dozens of artistic and literary creations, is another. Highwaymen like Dick Turpin were attractive figures and especially apt to be transformed into romantic heroes in nineteenth-century fiction.

Whether by first-hand experience, or exposure to the bodily relics of post-execution punishment and the associated retelling of remembered narratives, or in literary and other artistic creations of varied type and merit, bodies of criminals subject to post-execution punishment under the Murder Act had an impact in society that was disproportionate to their small numbers.

4. While anatomical dissection and hanging in chains are offered as equal alternatives by the Murder Act, no surviving written guidance is offered as to which one of the two should be specified under which circumstances. This suggests that the legislators behind the act perceived the two punishments to be equivalent. Nevertheless, the symbolic implications and historical context of the two alternatives are very different.

Hanging in chains traces a history through the medieval and early modern traditions of punishment that take their supposed deterrent and retributive effects from public humiliation of the body. As a punishment it emphasises the particular, unique and individual body of the malefactor, whose name is sometimes even immortalised by being written permanently into the landscape.

By contrast, anatomical dissection partakes not primarily in the discourse of punishment but that of science. The value of the criminal body to science is not in its particular history or its criminality but in its universality, its capacity to stand for the body of any human man or any human woman. Dissection as a mortuary treatment results not in fossilising it into its place, but rather, when carried out 'to the extremities' in the obliteration of the criminal self altogether. Anatomy belongs to the modern discourse of medical progress and scientific knowledge; gibbeting to the medieval punitive discourse of bodily retribution. In this context, Elizabeth Hurren's argument, stemming from her research on this project, that anatomisation and dissection are not the same thing is pertinent. Hurren suggests that 'anatomisation' was defined by penal surgeons in practical terms as the legal checking mechanism for declaring medical death by registering the cessation of activity in the heart and lungs, and later in the heart, lungs and brain. At this stage of punishment the identity of the individual was still important—it was a key step in seeing that justice had been done. This 'anatomisation' was part judicial punishment, part crowd-pleasing spectacle (it involved making a token cruciform cut to expose the body's interior, but did not include any detailed scientific study), and part ritual theatre. 'Dissection' then referred specifically to post-mortem exploration of the corpse—cutting 'to the extremities on the extremities' until the body was despoiled (less than one-third left). In other words, they are two separate punishment steps.

In order to posit anatomical dissection as an equivalent to hanging in chains, the former needs to be interpreted only as an act of violence, not as a technique of scientific research. Respectable, educated men of science are reduced in status to the level of brutal torturers. Intellectual and philanthropic motivations were ignored or wilfully misinterpreted and instead the anatomist was popularly represented as taking a personal delight in cutting and disfiguring dead bodies.

5. Post-execution punishments derive power from the manipulation of liminal spaces, both geographically and conceptually. Anthropologically, liminal places are places in between, places where transformation from one state to another occurs and which belong, therefore, neither wholly to one state nor to the other. Death, like birth and puberty, is a liminal life stage. Liminal places are also dangerous places because there is always a risk that the transformation is not successfully accomplished and the outcome is either that the wrong end result is achieved or that what is supposed to be an ephemeral transitional state is prolonged. Society develops rituals and processes by which liminality can be negotiated and the transformation completed successfully. An execution is a particularly controlled passage through liminality. The execution and the stages leading up to it and following afterwards is a highly orchestrated ritual, a choreography of passage from the state of being alive to the state of being dead; from a living human to a corpse.

Freud talked about the '*unheimlich*' or uncanny feelings that result from making the familiar into the strange.[48] Death and dissolution is supremely '*unheimlich*', says Freud, because the dead body falls into the deepest part of what his successors have called the 'uncanny valley' between living originals and wholly inanimate representations such as pictures. Julia Kristeva used the term 'abject' to mean something that falls outside or has been thrown out of the normal symbolic order.[49]

A newly dead person is abject and *unheimlich*: the familiar made strange; the responsive made unresponsive. Dead bodies are not alive, but may still look alive. They are simultaneously alarming and compelling; simultaneously specific and universal; they are one person and all people. Post-execution rituals of the Murder Act period organise the exposure of criminal bodies to maximise their liminal and uncanny power.

6. Contradictions are a necessary part of any attempt to capture attitudes to the criminal corpse. If the germ of our project arose from an awareness that different belief discourses around the dead body produced apparently irreconcilable 'true' beliefs, a close study of the criminal corpse has not facilitated the creation of a single or integrated narrative. There remain important contradictions, great variation and apparently contrasting tendencies that cannot be resolved. These significant issues include the different histories and trajectories of the two main post-mortem punishments of the period: anatomisation and dissection; and hanging in chains. Neither have we found a straightforward or universal answer to the question 'What did people think about punishment under the Murder Act?' We have discovered examples of those who were horrified and deeply frightened by the gibbet or the dissection table, and those who joked about them; those who believed such sanctions to be effective and necessary deterrents and those who found them barbaric and redundant. As the period progressed the general tendency was a move away from public display, a trajectory that eventually moved executions themselves behind prison walls, closed the doors of anatomy theatres and brought down the gibbets. But at the same time, the last two incidents of hanging in chains in England, both in August 1832, attracted enormous crowds: James Cook's gibbet had to be taken down by special order of the Home Office because the thousands of people who had thronged to the spectacle were obstructing traffic and constituted a threat to public order. So it seems there was no loss of

popular appetite for post-mortem punishment even when the rhetoric directed against it was ubiquitous in the press.

There are other contradictions. The inability of the research team to reconcile these contradictions and to provide a single, coherent narrative of post-mortem punishment is not a failure, however, but a recognition of the complex and incommensurable experiences and understandings of the time. There never was 'an eighteenth-century attitude to dissection', just as there was no eighteenth-century attitude to the body itself; and just as, indeed, there is no twenty-first-century attitude to either of those things. Beliefs and values varied according to social background, individual personality and the context of asking. Even a single person was—and is—capable of holding multiple, parallel beliefs which are drawn upon contextually. Context not only informs belief, it actively shapes attitudes. Distaste for dissection, while not entirely created by its punitive use, was certainly reinforced and shaped by its history as a judicial sanction for the worst of criminals. In the context of twentieth-century organ donation, by contrast, a narrative of sacrifice constructs a highly invasive intervention in the dead body as an act of nobility, rationality and selflessness.

7. Our research disrupts the conventional historical narrative of punishment as a steady progression away from brutal physical and retributive punishment towards humane, reformatory punishment. This is a progressivist and Whiggish kind of history which, until recent decades was unproblematically interpreted as part of the general Improvement of Western society. The contribution of Foucault was to interpret the same narrative in terms of power, particularly the subtle workings of state power. So the transformation from ostentatiously violent punishment to the reforming penitentiary is not a fundamental shift in attitudes to social deviance but just a new strategy for bringing about the same end: shaping a potentially disorderly people to a compliant population who will fall into line with the wishes of the state.

8. When is death? Attempts to pinpoint or define the timing of death are seriously undermined by the history of the executed body. Not only was there considerable uncertainty around the moment of actual, biological death, evident most clearly in Hurren's shocking discovery that maybe a third of 'dead' bodies that were delivered to the anatomist still had beating hearts, but saying clearly what makes

a person alive or dead turns out to be far from straightforward.[50] The numerous ways in which medically 'dead' bodies continue to perform the social actions and to be accorded the same relational status as the living was explored in a workshop held by the project team, subsequently edited for publication by Shane McCorristine.[51] A person can be socially dead long before their body stops metabolising; just as a person can form an import node in relationships long after it has stopped: what Hallam, Hockey and Howarth robustly call 'vegetables' and 'vampires' respectively.[52] As Thomas Laqueur remarked, in an observation that resonated deeply with the team, 'becoming really dead—even in the West, where supposedly death is a precipitous event—takes time'.[53]

NOTES

1. Tomasini, F. (2017), *Remembering and Disremembering the Dead: Posthumous Punishment, Harm and Redemption over Time* (London: Palgrave).
2. See, Feinberg, J. (1984), *Harm to Others* (New York: Oxford University Press); Pitcher, G. (1984), 'The Misfortunes of the Dead', *American Philosophical Quarterly*, Vol. 21, 183–188.
3. See, Tomasini, F. (2017), *Remembering and Disremembering the Dead: Posthumous Punishment, Harm and Redemption over Time* (London: Palgrave), quote at p. 13.
4. The organ retention scandals at Alder Hey and the Bristol Royal Infirmary led directly to the passage of the 2004 Human Tissue Act, which regulates the removal, storage and treatment of human tissue.
5. See, Corns, C. and Hughes-Wilson, J. (2002), *Blindfold and Alone: British Military Executions in the Great War* (London: Cassell), pp. 103–104.
6. Following the media attention in the early twenty-first century given to the campaign to pardon those 'shot at dawn', this group of unfortunate victims of history attained considerable significance and recognition in general public culture. Storylines in Downton Abbey and The Village, both popular television serials, featured individuals who had been shot for cowardice and the subsequent exclusion of their names from the local war memorial. The stigma of their fate was shown to impact on their families long after the end of the war.
7. Carr, E.H. (1961), *What Is History?* (Cambridge: University of Cambridge Press). See also, Appleby, J.O., Hunt, L.A., and Jacob, M.C. (1994), *Telling the Truth About History* (New York: W. W. Norton & Company).

8. Katharine Park has advanced the view that late medieval Italians had a different view of the dead body and were less troubled by post-mortem intervention. See, Park, K. (1994), 'The Criminal and the Saintly Body: Autopsy and Dissection in Renaissance Italy', *Renaissance Quarterly*, Vol. 47, Issue 1, 1–33.

9. Godfrey, B. (2016), 'The Crime Historian's Modi Operandi', in Knepper, P. and Johansen, A. eds. *The Oxford Handbook of the History of Crime and Criminal Justice* (Oxford: Oxford University Press), pp. 38–56, p. 51.

10. See for example, Pluciennik, M. (2001), 'Introduction. The Responsibilities of Archaeologists', in Pluciennik, M. ed., *The Responsibilities of Archaeologists: Archaeology and Ethics* (Oxford: Archaeopress), pp. 1–8.

11. See for example, Potter, P. (1991), 'What Is the Use of Plantation Archaeology?', *Historical Archaeology*, Vol. 25, Issue 3, 94–107.

12. See, Tarlow, S. (2001), 'The Responsibility of Representation', in Pluciennik, M. ed., *The Responsibilities of Archaeologists: Archaeology and Ethics* (Oxford: Archaeopress), pp. 57–64; Tarlow, S. (2006), 'Archaeological Ethics and the People of the Past', in Scarre, C. and Scarre, G. eds. *The Ethics of Archaeology: Philosophical Perspectives on Archaeological Practice* (Cambridge: Cambridge University Press), pp. 199–218; Tomasini, F. (2009), 'Is Post-mortem Harm Possible? Understanding Death Harm and Grief', *Bioethics*, Vol. 23, Issue 8, 441–449; Tomasini, F. (2017), *Remembering and Disremembering the Dead: Posthumous Punishment, Harm and Redemption over Time* (London: Palgrave).

13. See, Tarlow, S. (2001), 'The Responsibility of Representation', in Pluciennik, M. ed., *The Responsibilities of Archaeologists: Archaeology and Ethics* (Oxford: Archaeopress), pp. 57–64, quote at p. 62.

14. See, Richardson, R. (1987), *Death, Dissection and the Destitute* (London: Routledge and Kegan Paul), pp. 280–281.

15. Ibid.

16. See, Metcalf, P. and Huntingdon, R. (1991), *Celebrations of Death: The Anthropology of Mortuary Ritual* (Cambridge: Cambridge University Press).

17. See, Sillar, B. (1992), 'The Social Life of the Andean Dead', *Archaeological Review from Cambridge*, Vol. 11, Issue 1, 107–123.

18. See, Bloch, M. (1971), *Placing the Dead: Tombs, Ancestral Villages, and Kinship Organisation in Madagaskar* (Cambridge, MA: Academic Press).

19. 'Progress of Cremation in England & Wales, Scotland and N. Ireland, 1885–2014,' statistics compiled by the Cremation Society of Great Britain, http://www.srgw.info/CremSoc4/Stats/National/ProgressF.html (Accessed 6 September 2017).

20. Tomasini, F. (2017), *Remembering and Disremembering the Dead: Posthumous Punishment, Harm and Redemption over Time* (London: Palgrave).
21. This is in essence the line taken in the classic accounts of Douglas Hay and Peter Linebaugh. Hay, D., 'Property, Authority, and the Criminal Law'; Linebaugh, P. (1975), 'The Tyburn Riot Against the Surgeons,' in Douglas Hay et al. eds. *Albion's Fatal Tree: Crime and Society in Eighteenth-Century England* (New York), pp. 17–64 and 65–118.
22. See, Gattrell, V.A.C. (1994), *The Hanging Tree: Execution and the English People 1770–1868* (Oxford: Oxford University Press).
23. See, Sappol, M. (2002), *A Traffic of Dead Bodies: Anatomy and Embodied Social Identity in Nineteenth-Century America* (Oxford: Princeton University Press); Laqueur, T.W. (2015), *The Work of the Dead: A Cultural History of Mortal Remains* (Oxford: Princeton University Press).
24. See, Hurren, E.T. (2016), *Dissecting the Criminal Corpse: Staging Post-execution Punishment in Early Modern England* (Palgrave Macmillan).
25. This piece was published on the University's 'Think Leicester' blog at http://www2.le.ac.uk/offices/press/think-leicester/arts-and-culture/2016/the-enduring-imaginative-power-of-the-criminal-corpse.
26. Battell Lowman, E. and Tarlow, S. (2017), 'Le Gibbet Anglais', *Criminocorpus* [forthcoming].
27. Sappol, M. (2002), *A Traffic of Dead Bodies: Anatomy and Embodied Social Identity in Nineteenth-Century America* (Oxford: Princeton University Press), pp. 290–291.
28. Jordanova, L. (1989), *Sexual Visions: Images of Gender in Science and Medicine Between the Eighteenth and Twentieth Centuries* (Madison: University of Wisconsin Press).
29. See, Helen Macdonald (2006), *Human Remains: Dissection and Its Histories* (Yale University Press).
30. Kott, J. (1993), 'Why Did Antigone Kill Herself?', *New Theatre Quarterly*, Vol. 9, Issue 34, 107–109.
31. Roundell, J.A.E. (1884), *Cowdray: The History of a Great English House* (London: Reprinted by Fb &C Limited, 2016).
32. *The True and Illustrated Chronicles of the Last Man Gibbeted in Yorkshire* (1900) (reprinted by January Books).
33. See, McLynn, F. (2013), *Crime and Punishment in Eighteenth-Century England* (London: Routledge), p. 50.
34. See, King, P. (2003), 'Moral Panics and Violent Street Crime, 1750–2000: A Comparative Perspective', in Godfrey, B., Emsley, C., and Dunstall, G. eds. *Comparative Histories of Crime* (Cullompton: Willan), pp. 53–71; Emsley, C. (2008), 'Violent Crime in England in 1919:

Post-war Anxieties and Press Narratives', *Continuity and Change*, Vol. 23, Issue 1, 173–195, 174.

35. See, King, P. (2017), *Punishing the Criminal Corpse, 1700–1840* (Palgrave Macmillan).

36. *Leicester Chronicle*, 28 July 1832.

37. R.v. McKay and Lamb [1837], available at http://www.law.mq.edu.au/research/colonial_case_law/nsw/cases/case_index/1837/r_v_mckay_and_lamb/ (Accessed 28 June 2017).

38. See, Anderson, C. (2015), 'Execution and Its Aftermath in the Nineteenth-Century British Empire', in Ward, R. ed., *A Global History of Execution and the Criminal Corpse* (Palgrave Macmillan), pp. 170–198.

39. Foxhall, K. (2016), *Health, Medicine and the Sea* (Manchester: Manchester University Press).

40. Forbes, T.R. (1978), 'Coroner's Inquisitions on the Deaths of Prisoners in the Hulks at Portsmouth, England, in 1817–1827', *Journal of the History of Medicine and Allied Sciences*, Vol. 33, Issue 3, 356–366, quote at p. 358.

41. Campbell, C.F. (1994), *The Intolerable Hulks: British Shipboard Confinement 1776–1857* (Cirencester: Heritage Books), pp. 34–35.

42. Sharpe, J.A. (1990), *Judicial Punishment in England* (London: Faber & Faber), quote at p. 53.

43. King, P. (2000), *Crime, Justice and Discretion in England 1740–1820* (Oxford: Oxford University Press).

44. See, Tarlow, S. (2017), *The Golden and Ghoulish Age of the Gibbet in Britain* (Palgrave Macmillan); McCorristine, S. (2014), *William Corder and the Red Barn Murder* (Palgrave Macmillan).

45. See, Klass, D., Silverman, P.R., and Nickman, S.L. (1996), *Continuing Bonds: New Understandings of Grief* (Washington, DC: Taylor & Francis).

46. Poole, S. (2015), 'For the Benefit of Example: Crime-Scene Executions in England, 1720–1830', in Ward, R. ed., *A Global History of Execution and the Criminal Corpse* (Basingstoke: Palgrave), pp. 71–101.

47. King, P. and Ward, R. (2015), 'Rethinking the Bloody Code in Eighteenth-Century Britain: Capital Punishment at the Centre and on the Periphery', *Past and Present*, Vol. 228, 159–205.

48. Freud, S. (2003), *The Uncanny* (trans. David McLintock, London: Penguin Books). Originally published in German in 1919 as *Das Unheimlich*.

49. See, Kristeva, J. (1982), *Powers of Horror: An Essay on Abjection* (trans. Leon S. Roudiez, New York: Columbia University Press).

50. See, Hurren, E.T. (2016), *Dissecting the Criminal Corpse: Staging Post-execution Punishment in Early Modern England* (Palgrave Macmillan).

51. See, McCorristine, S. ed. (2017), *Interdisciplinary Perspectives on Death and Its Timing* (London: Palgrave).
52. See, Hallam, E., Howarth G., and Hockey, J. (2005), *Beyond the Body: Death and Social Identity* (London: Routledge).
53. See, Laqueur, T. (2011), 'The Deep Time of the Dead', *Social Research: An International Quarterly*, Vol. 78, Issue 3, 799–820.

INDEX

Printed by Printforce, United Kingdom